Recent Developments in Invertebrate Repellents

ACS SYMPOSIUM SERIES **1090**

Recent Developments in Invertebrate Repellents

Gretchen E. Paluch, Editor
EcoSMART Technologies, Inc.
Roswell, Georgia

Joel R. Coats, Editor
Department of Entomology
Iowa State University
Ames, Iowa

Sponsored by the
ACS Division of Agrochemicals

American Chemical Society, Washington, DC

Distributed in print by Oxford University Press, Inc.

Library of Congress Cataloging-in-Publication Data

Recent developments in invertebrate repellents / Gretchen E. Paluch, Joel R. Coats, editor[s] ; sponsored by the ACS Division of Agrochemicals.
p. cm. -- (ACS symposium series ; 1090)
Includes bibliographical references and index.
ISBN 978-0-8412-2675-3 (alk. paper)
1. Insect baits and repellents. 2. Invertebrates--Effect of pesticides on. I. Paluch, Gretchen E. II. Coats, Joel R. III. American Chemical Society. Division of Agrochemicals.
SB951.5.R43 2011
632′.62--dc23
2011046248

The paper used in this publication meets the minimum requirements of American National Standard for Information Sciences—Permanence of Paper for Printed Library Materials, ANSI Z39.48n1984.

Distributed in print by Oxford University Press, Inc.

PRINTED IN THE UNITED STATES OF AMERICA

Foreword

The ACS Symposium Series was first published in 1974 to provide a mechanism for publishing symposia quickly in book form. The purpose of the series is to publish timely, comprehensive books developed from the ACS sponsored symposia based on current scientific research. Occasionally, books are developed from symposia sponsored by other organizations when the topic is of keen interest to the chemistry audience.

Before agreeing to publish a book, the proposed table of contents is reviewed for appropriate and comprehensive coverage and for interest to the audience. Some papers may be excluded to better focus the book; others may be added to provide comprehensiveness. When appropriate, overview or introductory chapters are added. Drafts of chapters are peer-reviewed prior to final acceptance or rejection, and manuscripts are prepared in camera-ready format.

As a rule, only original research papers and original review papers are included in the volumes. Verbatim reproductions of previous published papers are not accepted.

ACS Books Department

Contents

Indexes

Preface

As the global climate changes, scientists anticipate that the distribution of animal populations and disease vectors will expand. In the case of arthropods, such efforts hold immense significance as they have the potential to increase human mortality and suffering from arboviruses above current levels. The 238th American Chemical Society National Meeting and Exposition in Washington, D.C. on August 16-20, 2009, offered an opportunity for researchers to present and discuss new findings in invertebrate repellents research, regulations, and technology development.

Recently efforts have been made to understand the role of chemicals in arthropod behavior, and screening programs are starting to incorporate repellency testing into their battery of bioassays.The lack of standardized protocols for measuring and comparison of repellents has remained a significant obstacle in arthropod research. Oftentimes studies report variable measures of success, and comparison of results across studies is not always consistent. Progress in the standardization of arthropod test methods for repellents would be valuable to many groups including academic researchers working in the field, contract labs supplying test results, government research laboratories, regulatory bodies in the process of developing guidelines for product registration, as well as companies looking to invest in new technologies. Perhaps one complicating factor in this process has been that research and technology haven't moved fast enough to meet the demand for effective arthropod repellents. Issues such as pest arthropod resurgence and insecticide/repellent resistance to chemical can create new challenges and add pressure for researchers.

The collection of chapters in this book covers a range of applied and basic research on arthropod repellents. An overview of the state of arthropod repellents research is provided at the start. In the chapters that follow, there is a selection of papers demonstrating research on new repellent technologies at different stages of development. The scope of basic and applied research methods described in these chapters on new repellent technologies presents the range of testing that is often necessary to move a repellent technology forward in development. The transition from newly developed technologies to registered products is achieved in perspective of a growing markert for natural arthropod repellents. New technologies that are completely developed and have gone through registration need to be accompanied by successful commercialization. The growing market for natural arthropod repellents presents such an example and highlights new opportunities in this area. The concluding chapter discusses the public entomology landscape, past and future opportunities for the development of chemical protectants.

Our intent with this ACS Symposium Series book is to summarize the current state of arthropod repellent research, draw connections between basic and applied studies, highlight the need for standardized test methods, provide information on the framework for product registration, and encourage future investments in technology by example of a growing market segment.

We greatly appreciate the efforts put forth by the contributing authors, session presenters, reviewers, and staff members at ACS Books. We dedicate this book to family and friends (by GP) and to all of my children and grandchildren (by JC).

Gretchen E. Paluch
Director of Basic Research
EcoSMART Technologies, Inc.
Roswell, Georgia

Joel R. Coats
Distinguished Professor
Department of Entomology
Iowa State University
Ames, Iowa

Chapter 1

A Review of Arthropod Repellents

Marc C. Dolan*[,1] and Nicholas A. Panella[2]

[1]Centers for Disease Control and Prevention, Division of Vector-Borne Diseases, Bacterial Disease Branch, 3150 Rampart Road, Fort Collins, Colorado 80521
[2]Centers for Disease Control and Prevention, Division of Vector-Borne Diseases, Arboviral Disease Branch, 3150 Rampart Road, Fort Collins, Colorado 80521
***E-mail: mcd4@cdc.gov**

Arthropod bites can potentially result in the transmission of numerous infectious diseases and remain a leading cause of human morbidity and mortality worldwide. The most effective means of preventing arthropod bites is achieved through the use and practice of personal protective measures, including the use of repellents. Repellents are typically applied to exposed skin but they can also be applied to clothing or other surfaces to discourage arthropods from landing or climbing onto treated surfaces. In this chapter we review the history of repellents, how we attract biting arthropods, and provide some detail on how repellents work. Information is provided on the effectiveness of four common synthetic compounds including: Deet, permethrin, picaridin, and IR3535. In addition, efficacy of naturally derived repellents such as: citronella, lemon eucalyptus oil, BioUD and other all natural compounds are discussed. Finally, current research on novel all natural compounds are reported.

Introduction

Arthropod bites can potentially result in the transmission of numerous infectious diseases and remain a leading cause of human morbidity and mortality worldwide (*1*). Mosquitoes transmit disease to more than 700 million people annually (*2*) and mosquito-borne malaria alone kills 3 million people each year (*3*). Although arthropod-borne diseases are usually associated with tropical and subtropical regions of the world, vector-borne infectious diseases remain a significant threat in temperate zones of the Unites States (US). Mosquito-borne diseases in the US such as Eastern equine encephalitis virus (EEE), West equine encephalitis virus (WEE), St. Louis encephalitis virus and La Crosse encephalitis virus (LACV) result in hundreds of clinical cases annually. West Nile virus, first discovered in 1999 in New York City, has been responsible for disease outbreaks of epidemic proportion leading to several thousand neuro-invasive and fatal cases over the past decade. Additionally, public health officials are reporting an increase in the number of cases of yellow fever and other hemorrhagic fevers (dengue) along the US/Mexico border (*4*, *5*).

In addition to mosquitoes, fleas and ticks are also important vectors of infectious disease in the US. Fleas serve as the primary vector of plague (*6*), and ticks transmit more disease-causing organisms than any other hematophagous arthropod. These include the agents that cause Lyme disease, *Babesia, erhlichiosis* (*7–9*), as well as Rocky Mountain spotted fever, relapsing fever, and Colorado tick fever (*10*, *11*).

Control of arthropods is typically achieved through the use of chemical pesticides. While area-wide applications of insecticides has been shown to be effective at reducing medically important species of mosquitoes, ticks, and fleas, there are an ever growing number of problems associated with pesticide use (*12*, *13*) including: environmental contamination, impact on non-target organisms, persistence in the environment, development of resistance, and expense (*14*). The most effective means of preventing arthropod bites is achieved through the use and practice of personal protective measures which include: avoidance of arthropod habitat, wearing protective clothing, limiting outdoor activity during periods of highest risk (dusk and dawn for mosquitoes), alteration of landscape to reduce acceptable arthropod habitat, and the use of repellents. An alternative to pesticides, an arthropod repellent can be used as an easy and effective choice to reduce or eliminate the risk of acquiring these diseases (*4*). Repellents are typically applied to the skin to protect against biting arthropods. Repellents can also be applied to clothing or other surfaces which discourage arthropods from landing or climbing onto treated surfaces (*4*).

A repellent can generally be described as a substance that can be used “to cause movement away from a stimuli”, “to be repulsed” or “an agent of action” as in “any stimulus which elicits an avoiding reaction” (*1*, *2*, *15*, *16*). Repellents are available to consumers in a variety of products with a multitude of formulations and applications. Examples include: aerosols, pump sprays, lotions, creams, sunscreen sprays and creams, towellettes, powders, grease sticks, impregnated wrist bands, and impregnated clothing materials (*15*, *16*). An ideal repellent should provide protection for up to 8 hours against an array of blood-feeding arthropods with a single treatment, be safe for application to skin and clothing, non-toxic to adults and children alike, be virtually odorless, non-greasy, have no effect on clothing fibers such as staining or bleaching, and be economical. The search for such an arthropod repellent continues (*17*, *18*). Research efforts to discover the ideal repellent are often hampered by the many variables inherent to the repellency of synthetic and natural compounds alike. Like pesticides, repellent compounds do not share common modes of action. In fact, very little is known about the intrinsic mode of action of repellents and how they repel target arthropods.

All repellents exhibit some degree of volatility and when repellents are applied to either skin or clothing, it allows for the production of a vapor layer, creating an unpleasant or offensive surface, smell, or taste to biting arthropods (*19*). All repellent compounds have a relative vapor pressure which is directly correlated to vapor repellency. When vapor repellency is correlated with the boiling point of the chemical compound, optimal effective range falls somewhere between 230 to 260°C, meaning that compounds with those boiling points have enough volatility to exert some vapor repellency, but not so much volatility that they evaporate away quickly (*15*). Therefore, synthetic chemicals and naturally derived compounds with high vapor pressures will dissipate rather rapidly whereas those with low vapor pressures will vaporize too slowly and may not supply enough volatile repellent compound to be effective (*15*, *20*, *21*). The mode of action for most repellents occurs by forming a repellent barrier that resides within one inch of the treated surface area. Rather than camouflaging the human body's attractants (heat, CO_2, lactic acid), they cause biting arthropods to turn away as they approach the repellent barrier. This means that a repellent applied to the back of the hand will not protect the palm of the hand or forearm from biting arthropods. A repellent's efficacy can be dramatically affected by sweating, abrasion of treated areas, heat, humidity, getting treated areas wet, and washing with soap and water. In addition, environmental factors such as temperature, wind, and humidity can affect repellent delivery systems, thereby influencing repellent effectiveness by impacting variability (*15*). In some cases less than 1% of active ingredient can form a repellent barrier, but most commercial formulations include higher percentages of active ingredient. A repellent's effectiveness is a combination of the relative vapor pressure (volatility) and delivery system (formulation). This combination will determine how much and how often one must apply a given repellent in order for it to be effective. Different repellents will require different levels of application and re-application and one must also consider the type of activity that will be endured and the type or types of insects you are trying to avoid. Another consideration is the disease risk associated with certain biting arthropods.

The most significant research regarding the discovery of novel repellents has been conducted by the US military in order to protect troops from arthropod-borne pathogens. The most important discovery to emerge from the military research program is Deet (N,N-diethyl-m-toluamide). Deet has been the most extensively used repellent for nearly 60 years. It repels a broad spectrum of biting arthropods and is available in many different concentrations in a number of application products. Deet has been described by some to have a foul odor and an oily greasy feel. Deet has also been labeled as a plasticizer (capable of dissolving watch crystals, plastics, and certain clothing fibers), and adverse health effects have been reported by some, but the number of cases is low compared to the number of applications. Despite the long successful history of Deet, contemporary research programs have focused on the development of alternative repellent compounds. The Centers for Disease Control and Prevention (CDC) recommends three alternatives to Deet to repel mosquitoes and ticks: oil of lemon eucalyptus (p-menthane- 3, 8 diol or PMD), IR3535 (3-[N-butyl-N-acetyl]-aminopropionic acid, ethyl ester) and a piperidine, picaridin (2-(2-hydroxyethyl)-1-piperidinecarboxylic acid 1-methylpropyl ester) (*23*). Of the three mentioned above, only PMD is a naturally derived product, and research in that direction has increased significantly in recent years bringing more naturally derived compounds to the repellent market. It is important to note that several factors have contributed to an increase in human exposures to insect-borne pathogens. Among the most important is the continued expansion of human populations from urban to rural areas, an increase in international travel, and the emergence of novel vector-borne infectious diseases (*1*, *5*, *22*, *23*). The development of novel botanical-based repellents is crucial as an increasing proportion of the human population world-wide chooses not to use Deet or synthetically produced repellent products.

How We Attract Biting Arthropods

Factors involved in attracting biting arthropods to a host are numerous, complex, and not fully understood. Mosquitoes and other flying insects rely on visual, thermal, tactile, and olfactory cues to locate a potential host. Olfactory cues are believed to play the most vital role in attracting mosquitoes (*17*). Different species of mosquitoes target various hosts and therefore, may be active at different times. Diurnal species for instance appear to rely heavily on visual clues including movement and color, (they tend to be attracted to darker colored fabrics). Olfactory cues are most important when a mosquito has located a host and is within feeding range (*17*). It has been estimated that the human body produces nearly 400 detectable compounds as byproducts from metabolism and greater than 100 volatile compounds in the human breath (*24*). Carbon dioxide and lactic acid are two of the best-studied mosquito attractants (*17*). Mosquitoes use chemo-receptors located on their antennae to detect these compounds. At close range, skin temperature and moisture in the form of sweat may also further attract mosquitoes and other biting arthropods (*24*, *25*). Biting arthropods may show an affinity for certain parts of the human body that vary even among species. Some individuals may be more attractive to biting arthropods than others due to variations in cues or whole body odors which seem to be more attractive than any single cue acting alone. Certain lotions, soaps, and perfumes may also be attractive to biting arthropods. Adults tend to be more attractive than children, men more so than women, and larger individuals tend to receive more bites than others, most likely due to higher amounts of heat, CO_2 and lactic acid secretions (*17*, *26*).

Unlike flying pests, crawling, terrestrial arthropods and ticks in particular, will locate suitable hosts by ambushing, hunting, or a combination of the two. Most tick species tend to climb up vegetation with their forelegs extended and wait for a host to come by. This behavior is referred to as questing. Ticks have sensory organs located on the tarsi of the front legs. Ticks that actively hunt or ambush hosts utilize stimuli such as carbon dioxide, lactic acid, heat, and vibration (*18*). Relatively little research has been conducted to determine repellent mode of action on ticks. Most published repellent assays utilize vertical, horizontal, or treated finger bioassays to determine repellent efficacy (*27*, *28*). Unfortunately, these types of assays do not necessarily discriminate between repellency due to olfaction or tactile chemoreception (*29*).

History of Arthropod Repellents

Repellents can be traced back thousands of years when our early ancestors used tars, smokes, various plant oils, soils, and other methods (*30*). Smoke was and continues to be the most widely used repellent means for mosquitoes in tropic and subtropical regions of the world (*31*). While effective, the use of smoke requires continuous production resulting in poor residual activity (*32*). Safer and more modern methods of repelling mosquitoes, including personal repellents, were needed. Prior to World War II, only four primary personal repellents existed. The most widely used prior to 1940 was oil of citronella. This compound was discovered in 1901 and was primarily used as a topical for fleas and head lice but is still widely used today in various formulations. The other 3 main repellents include dimethyl phthalate, discovered in 1929, Indalone®, patented in 1937, and Rutgers 612, which was made available to the public in 1939 (*15*, *30*). In 1953, the insect repellent Deet was discovered, and the first Deet containing products were introduced in 1956. Since its introduction, Deet has been considered the most efficacious and most used arthropod repellent. Several other compounds and thousands of natural based compounds have been researched and evaluated but none have enjoyed the success of Deet (*15–18*)

Compounds

Deet

A breakthrough in repellent history occurred when the U.S. Department of Agriculture discovered a compound which was later patented by the U.S. Army. Previously called N,N-diethyl-m-toluamide, N,N,-diethyl-3-methylbenzamide (Deet) remains the most widely used repellent on the market for the last 50+ years and remains the gold standard even today. Deet is considered the most broad-spectrum, efficacious arthropod repellent ever produced. It has been shown to be effective against all species of mosquitoes, *Aedes spp*, (*33*, *34*), *Culex spp.* (*35*), and *Anopheles spp.* malaria vectors (*36*). In addition, Deet repels other biting insects including sand flies (*39*), as well as ticks (*37*), and chiggers (*38*).

There are an estimated 140 products containing Deet produced by 39 companies registered with the EPA (*40*). Deet is available in products with concentrations ranging from 5 - 95% with a majority of products containing ≤ 35%. The Centers for Disease Control (CDC) recommends using products that contain < 50% Deet as the duration of activity does not increase with increased concentrations of active ingredient above 50% (*23*). As an example, in a laboratory study, 50% Deet provided 4 hours of protection from the bite of *Ae aegypti* while 100% Deet provided 5 hours (*41*). Some of the most popular consumer products containing Deet are Deep Woods Off® and Family Care® brands (SC Johnson, Racine, WI), Cutter Backwoods® (Spectrum Brands, Atlanta, GA), and Ultrathon® (3M, St. Paul, MN).

Deet is designed to be applied to exposed human skin and clothing and repels insects as opposed to killing them. While the application of Deet to certain polyester and cotton fabrics appears to increase repellency, Deet is also a known plasticizer and may damage certain fibers including polyester as well as watch crystals. In addition, many consumers dislike the feel and odor of Deet products and question its safety. In 1998, EPA issued new labeling standards and manufacturers of Deet-containing products could no longer claim the product as child safe; although Deet can be used safely on children when used according to the label (*40*). The American Academy of Pediatrics (AAP) released their recommendations for use of Deet products on children in 2003: insect repellents containing Deet with concentrations ranging from 10 – 30% appear to be safe products when used in accordance with product labels (*23*). As mentioned previously, the number of adverse events in relation to the number of applications is extremely small, and with proper use and adherence to safety labels, this makes Deet one of the best broad spectrum repellents available (*42*).

Permethrin

Permethrin, [3-(phenoxyphenyl) methyl (±)-*cis*, *trans*-3-(2,2-dichloroethenyl)-2,2-dimethyl-cyclopropanecarboxylate], is a synthetic pyrethroid that was designed to mimic natural pyrethrins which are derived from the crushed dried flowers of the chrysanthemum plant (*42*, *43*). Permethrin is unique in the fact that it functions as both a contact insecticide and a repellent and is active against a wide variety of biting arthropods. The primary mode of action of permethrin is it's binding at the sodium channel receptor sites in a way that prevents the complete closing of the sodium channel, resulting in sustained slow leakage of sodium ions into the neuron (*44*). Permethrin was first marketed in 1973 and has not only been used as a repellent but widely used as an agricultural, forestry, home pest control and public health pesticide (*15*, *16*).

Because permethrin is synthetically derived and functions as a contact insecticide, it is not safe for application to human skin but rather as a clothing treatment. Permethrin has proven to be extremely effective as a repellent when applied to clothing for personal protection against many biting arthropods (*45*, *46*). Applications made to clothing can last multiple washings (*47*). In addition to clothing, permethrin can be applied to mosquito nets, curtains, tents, and blankets

(*48*). One of the more notable campaigns is the use of permethrin-treated bed nets in malaria endemic areas. This has proven to be a very effective and affordable method to reduce vector transmission of medically important diseases (*42*).

Permethrin products are produced as gear and clothing treatment only and typically contain 0.5% active ingredient. Brands include Permanone® (Bayer, Pittsburgh, PA), Repel® brand and Coleman® Insect Treatment (WPC Brands, Inc, Jackson, WI), and Duranon® Tick Repellent (Sawyer Products). Products should be applied until material is moist and allowed to dry completely prior to use (≥ 2 hour). Permethrin is non-staining, non-greasy, and virtually odorless. Current formulations are UV-resistant and will typically last ≥ 2 week after a single application.

Picaridin (KBR 3023)

Picaridin [2-(2-hydroxyethyl)-1-piperidinecarboxylic acid 1-methylpropyl ester] is a relatively recently approved insect repellent in the United States. Picaridin contains one of the most common active ingredients in insect repellents approved for use in Europe and Australia where it is known by the trade names Bayrepel (KBR 3023, Bayer AG) and Autan (S.C. Johnson and Sons Inc.) respectively . Bayrepel was developed by Bayer AG who began researching for a new active ingredient in the early 1980's. Picaridin-containing products have many desirable characteristics that the public may perceive as the ideal personal repellent. Unlike Deet, picaridin is virtually odorless, does not have a greasy or sticky feel during or after application, will not damage fabrics, and is not a plasticizer (*16*, *40*, *42*). The mode of action of picaridin is not fully understood but it appears to provide a vapor barrier that deters biting insects, similar to Deet (*49*).

Picaridin was first used in Europe in 2001 and was registered in the US in 2005 (*40*). Field trials have shown picaridin to be effective against numerous species of mosquitoes, biting flies, and ticks. A field trial published by Barnard and colleagues demonstrated that a 25% formulation of KBR 3023 was nearly as effective as 25% Deet in preventing bites by *Ochlerotatus triseriatus* in the Florida Everglades (*34*). In 2001, Consumer Reports (*50*) reported that a 7% and 20% solution used by the Australian Army were effective (*51*, *52*). However, retests performed by Consumer Reports in 2006 showed that picaridin offered little protection against *Aedes* mosquitoes and a protection time of approximately 2.5 hour against *Culex* species (*53*). Laboratory tests performed in 2004 demonstrated that 10% KBR3023 was as effective as 15% Deet while offering protection times of 4-8 hour against 3 species of mosquitoes (*35*). Pretorius and others demonstrated that picaridin had a protection time of only 1 hour against *Amblyomma hebraeum* ticks while Deet provided protection for 2 hours (*24*).

In 2000, the World Health Organization (WHO) announced that picaridin was their recommended product for repelling *Anopheles* mosquitoes, the primary vector of malaria, due to its safety, effectiveness, and cosmetic properties (*54*). In 2005 the Centers for Disease Control and Prevention (CDC) added picaridin as a recommended active ingredient for preventing the transmission of West Nile

Virus (*23*). To date, picaridin has not been as extensively tested as Deet but is recommended by several authorities as a safe, effective, and pleasant alternative. Studies demonstrate that picaridin seems to be most effective at concentrations ≥ 20% (*55, 56*). The use of products with lower concentrations may require more frequent reapplication. In the US, products containing picaridin and marketed under the name Cutter Advanced® (Spectrum, St. Louis, MO) are available in pump formulations containing 7 and 15% active ingredient, Avon SSS Bug Guard + Insect Repellent® with 10% Picaridin (Avon Products, New York, NY), and Natrapel® 8 hour insect repellent with 20% picaridin (Tender Corporation, Littleton, NH).

Citronella

Citronella is an essential oil derived from lemongrass of the genus *Cymbopogon* comprising some 55 species of grasses. *Cymbopogon nardus* and *Cymbopogon winterianus* are the two species most commonly used to produce citronella oil for the food and insect repellent industries. The active compounds in citronella oil for repelling mosquitoes are camphor, eucalyptol, eugenol, linalool, citronellal and citral (*57*). The United States has recognized the use of citronella as an insect repellent since 1948 (*58*). Citronella is considered a biopesticide with a non-toxic mode of action by the United States Environmental Protection Agency. Numerous (161) published scientific studies on the efficacy of citronella oil as an insect repellent have been conducted, and a recently published review of 11 such studies can be found in an article published by Kongkaew and colleagues (*59*). The 11 studies that met the criteria proposed by the authors concluded that citronella, by itself, was not as effective as Deet (N,N-diethyl-m-toluamide) in terms of mean protection time against mosquito bites. In studies using *Aedes* species of mosquitoes the duration of mean protection time was significantly lower than that of Deet (*2*, *34*, *60*). However, citronella was considerably more effective at repelling mosquitoes of the genera *Anopheles* and *Culex*, sometimes outperforming Deet (*35*, *61–63*). Additionally, they also concluded that some formulations with vanillin increased protection time significantly. For instance, a formulation of 25% citronella combined with 5% vanillin increased complete protection time against *Anopheles* mosquitoes from 3 hours to 6 hours as compared to a 25% citronella only formulation (*60*).

In controlled laboratory studies, citronella demonstrated adequate repellency activity against *Aedes* mosquitoes and nearly equivalent protection as Deet against *Anopheles* and *Culex* mosquitoes. Real world studies may provide different results as mosquito species composition in the environment varies greatly. To ensure maximum protection against mosquito bites if using citronella as the primary repellent product, it should be applied every 30-60 minutes. Brands include: All Terrain Herbal Armor Insect Repellent® (Sunapee, NH), Natrapel Insect Repellent® (Tender Corporation, Littleton, NH), Buzz Away® (HOMS Inc, Pittsboro, NC), and Burt's Bees Outdoor All Natural Herbal Insect Repellent® (Durham, NC). Clearly, the decision to recommend citronella as a potential personal insect repellent with reasonably few adverse effects by the USEPA and the Centers for Disease Control and Prevention was warranted (*22*).

IR3535 (Avon Skin So Soft)

IR3535 (ethyl butyl acetyl aminopropionate) is an insect repellent found in Avon's (New York, NY) Skin So Soft® line of products. IR3535 is a synthetic compound that is structurally similar to naturally occurring β-alanine and is registered as both an insect repellent and a biopesticide by the USEPA (*17, 64*). IR3535 has been used as an insect repellent in Europe for over 20 years and was introduced to the US market in 1999 (*65*). Skin So Soft® originally received attention in the US when consumers began reporting on repellent effects of Avon's bath oil product. Initial tests demonstrated that Skin So Soft® oil provided only limited protection of ≤ 40 minutes (*17, 33*). Avon currently markets its products under the Skin So Soft® brand that contains the EPA-recognized repellent Picaridin.

Compared to Deet there are minimal published scientific studies measuring the repellent efficacy of IR3535. A majority of studies to date report that IR3535 moderately repels arthropods compared to other repellent active ingredients such as Deet, picaridin, and p-menthane 3,8-diol (oil of lemon eucalyptus). Laboratory studies performed against sand flies and black flies demonstrated repellency ranging from 5.9 – 10.4 hours (*66*). Additional laboratory studies targeting *Aedes* and *Culex* mosquitoes resulted in average protection times of 3.2 hours with a 7.5% formulation (*35, 67*). Three field trials tested IR3535 against mosquitoes and indicated that IR3535 was as effective as Deet in repelling *Aedes* and *Culex* mosquitoes while less effective than Deet in repelling *Anopheles* spp (*34, 36, 68*). A 2008 study evaluated time-release formulations of IR3535 and reported protection times from 7.1 – 10.3 hours for mosquitoes and 9.1 – 12.2 hours for blacklegged ticks (*68*).

IR3535 is safe to apply to both skin and clothing and appears to have an unblemished safety record. Evidence suggests that IR3535 will repel mosquitoes, ticks, chiggers, sand flies and biting midges. In 2009 the Centers for Disease Control and Prevention (CDC) added IR3535 as a recommended biopesticide repellent for preventing the transmission of West Nile Virus. IR3535 is available from Avon as Skin-So-Soft Bug Guard Plus Picaridin Insect Repellent Spray® and contains both IR3535 and 10% Picaridin (Avon Products Inc, New York, NY).

Natural Botanical Repellents

Botanical-based repellents typically contain one or more plant essential oil or compounds derived from essential oils. These include previously mentioned oil of lemon eucalyptus and citronella. Thousands of plants have been screened for repellent and insecticidal activity. Although naturally derived repellents have not been shown to be as efficacious as their synthetic counterparts; they may have a distinct advantage as being perceived as safer for use with less harmful side effects. One of the more recent natural repellents to enter the market is Organic Bite Blocker Extreme Insect Repellent® (HOMS, LLC, Clayton, NC). Bite Blocker Extreme® lists soybean, geranium, and castor oils as its active ingredients. A field trial conducted in Ontario, Canada reported that soybean oil provided 97% protection against Aedes mosquitoes after 3.5 hours (*17*, *69*). Another product, Burt's Bees All Natural Herbal Insect Repellent® (Burt's Bees, Inc, Durham, NC), contains a milieu of 8 all natural ingredients: castor, rosemary, lemongrass, cedar, peppermint, citronella, clove, and geranium oils. EcoSMART Organic Insect Repellent® (EcoSMART Technologies, Inc, Alpharetta, GA) contains rosemary, cinnamon, lemongrass and geranial and claims to repel mosquitoes, ticks, and gnats for hours.

Some of the more widely studied and effective oils include: thyme, geraniol, clove, and cedar oils (*70*, *71*). Laboratory trials conducted at Iowa State University described the repellent effects of extracts from Osage orange on the German cockroach (*Blattella germanica*) and the maize weevil (*Sitophilus zeamais*). In addition, they isolated the active ingredient in catnip, nepetalactone, and found it's isomers to be more effective by vapor repellency than Deet (*30*). Exploration of the plant kingdom will likely continue in the quest to discover safer alternatives to synthetic compounds such as Deet and permethrin. To date however, a majority of essential oils tend to give minimal protection, usually ≤ 2 hours. This may be attributed to the fact that most plant derived oils are highly volatile and UV-sensitive. However, expert formulations with more efficient carriers may be able to overcome the shortfalls attributed to essential oils.

PMD - Lemon Eucalyptus Oil

Known in the United States as oil of lemon eucalyptus (OLE) or under its trade name of Citriodiol, p-menthane- 3, 8 diol (PMD) is the active ingredient now found in many insect repellents. Long used as an ingredient in throat lozenges to ease sore throats, PMD along with citronella, is now recognized by the Centers for Disease Control and Prevention (CDC) to be the only effective naturally derived substance for deterring mosquitoes that transmit West Nile virus (*72*). PMD was first isolated from the "lemon scented gum" *Corymbia citriodora*, as the Australians call it, in a mass screening campaign of plants undertaken by the Chinese government beginning in 1960 to discover potential new insect repellents (*73*). Ironically, PMD is not the essential oil of eucalyptus, but a waste material originating from the hydrodistillation of the essential oil from the leaves (*74*).

Long used in China as a commercially available repellent the early testing on PMD in laboratories in the west showed mediocre to good repellent performance when compared to N, N-diethyl-m-toluamide (Deet) against different genera of mosquitoes (*75*, *76*). These early studies used the formulations obtained from Chinese producers that carried the active ingredient in ethanol most likely compromising its repellent qualities by evaporating quickly. In reformulations in the United Kingdom a few years later, PMD concentration was increased to 50% and ethanol was replaced with more cosmetically sophisticated carriers (*77*). These changes transformed PMD into a much more effective repellent. Several field investigations against *Anopheles* mosquitoes in Africa demonstrated that products containing this new formulation offered the same level of protection as repellents containing 50% Deet. Laboratory results were similar (*77*). Further research confirmed the initial results. For example: Barnard and Xue (2004) (*35*) ranked PMD first in a study examining 12 commercially available repellent products, some of which contained up to 30% Deet.

In a very comprehensive study comparing various concentrations of PMD to Deet, researchers found virtually no difference in repellent performance during a 6-hour field trial (*72*). In addition to repelling mosquitoes PMD has been shown to be an effective repellent of ixodid ticks in the laboratory ((*77*), *Dolan et al. unpublished data*). Given all the empirical data on the repellent qualities of PMD, it was likely a wise decision for the US government to acknowledge and recommend products with this agent to the general public.

2-Undecanone (BioUD)

The latest compound available to consumers for personal protection against tick and mosquito bites is 2-undecanone. Originally derived from wild tomato plant *Lycopersicon hirsutum* Dunal f. *glabratum* tissues, 2-undecanone is a known natural plant defense mechanism against insect herbivory (*78*) prompting investigators to experiment with its use as a topical insect repellent for humans. A methyl ketone, 2-undecanone is the active ingredient (7.75%) in the latest arthropod repellent registered for use by the USEPA: BioUD® (HOMS LLC, Clayton, NC). Published results of laboratory and field trials have used this formulation for comparisons to N, N-diethyl-3-methylbenzamide (Deet) and other commercially available products (*79*, *80*).

Using mosquito arm-in-cage studies researchers compared BioUD (7.75% 2-undecanone) against two Deet formulations of 7 and 15% respectively using *Aedes albopictus* and *Aedes aegypti* mosquitoes. Measurements of repellency were observed for 1-6 hours. BioUD was found to be equally effective as 7% Deet and nearly as effective as the 15% concentration of Deet in trials using *Aedes albopictus*, although these differences were not statistically significant (*81*). In trials using *Aedes aegypti* they found that BioUD was as effective as the 7% Deet formulation, but significantly less effective than the 15% DEET over the 6-hour trial period. Furthermore, using the same formulations in field evaluations against wild mosquito populations BioUD significantly outperformed BiteBlocker® (3% soybean oil, 6% geranium oil and 8% castor oil) as well as a 30% commercially available formulation of Deet (*81*). In this same study BioUD with 7.75% 2-undecanone was also shown to provide considerable repellency activity against ixodid ticks in both laboratory and field settings. Although initial laboratory and field trials show this to be a promising compound, further comparisons will have to be made in order for this compound to be recommended by public health authorities.

Human vector-borne diseases are a growing worldwide concern. In the US, ticks transmit Lyme disease, mosquitoes transmit West Nile virus, and fleas transmit the plague bacteria. Very few can be prevented with vaccines, and many are untreatable or unresponsive to antibiotics. The ability to effectively kill and repel vectors is the only means currently available to reduce disease risk. However, cost, environmental impact, insecticide resistance, and public concern all limit the usefulness of currently available synthetic pesticides and repellents. The need to discover alternatives to synthetic pesticides and repellents that are environmentally friendly and safe for human use has led scientists to explore products that can be developed from botanical sources.

Over the past 15 years at the Centers for Disease Control and Prevention, Division of Vector-Borne Disease, researchers have been investigating naturally derived products as both pesticides and repellents for controlling medically important arthropods. Investigations have focused on natural products derived from the essential oil of Alaska yellow cedar (*Chamaecyparis nootkatensis*). Laboratory bioassays were conducted to determine the activity of 15 chemical constituents isolated from the essential oil of Alaska yellow cedar (AYC) against *Aedes aegypti* mosquitoes, *I. scapularis* ticks, and *Xenopsylla cheopis* fleas. The compound nootkatone was found to be one of the most effective biocidal compounds with a mean LC_{50} range of 0.0029 – 0.0083% against ticks, fleas, and mosquitoes (*82*). Field trials conducted in a Lyme-endemic area of New Jersey demonstrated that a single area-wide application of 2% nootkatone controlled nymphal deer ticks at levels $\geq$ 91.6% for 42 days (*83*).

Initial efforts to determine the repellent efficacy of nootkatone were evaluated against nymphal deer ticks using a vertical laboratory bioassay and compared to technical grade Deet. Four hours after treatment, nootkatone had a repellent concentration (RC_{50}) value of 0.0458% (wt:vol) as compared to 0.0728% for Deet. Although the observed RC_{50} value was not statistically significant, the ability of nootkatone to repel ticks at relatively low concentrations may represent a safe alternative to Deet and permethrin (*84*). In repellent field trials using treated coveralls, ticks drags, and white cotton sheets, nootkatone was more effective at repelling both deer ticks and lone star ticks than both Repel® brand Permanone® (0.05% permethrin) and EcoSMART Organic Insect Repellent® (*85*, *86*).

Currently, all the compounds that CDC scientists are researching are natural and some, like nooktatone, are considered food-grade and are used as flavor and fragrance additives in the food and cosmetic industries. Nootkatone is essentially the essence of grapefruit and has a very pleasant, citrus-like odor. Equally important as its safety record, scientists at CDC and Iowa State University have demonstrated that nootkatone and the other compounds from AYC have a unique mode of action as compared to that of other known pesticides and repellents. These unique characteristics and attributes make them a potentially important alternative weapon against arthropods that have developed resistance to currently registered pesticides (*87*). Moreover, a large percentage of survey respondents claim that they would be more likely to use naturally derived insecticides and repellents than synthetics (*14*).

Conclusions

As discussed there are a number of USEPA registered repellents currently approved for personal use. The CDC recommends the use of Deet, picaridin, oil of lemon eucalyptus [active ingredient: p-menthane 3,8-diol (PMD)], and IR3535 (*23*). Deet continues to remain the gold standard to which all other repellents are measured against, due in large part to the wealth of scientific evidence indicating that this product has the longest duration of protection against the greatest spectrum of arthropods. While Deet has been used for nearly 6 decades with few adverse health effects (*88*), there continues to be consistent concern regarding its safety (*89*). Future scientific studies will continue to provide insight into the mode of action of repellents and which are most effective. Thousands of novel synthetic products and essential oils are screened for efficacy every year resulting in a broad range of repellents from which to choose. When choosing a repellent it is therefore important for the consumer to consider many factors including: active ingredient, concentration, rate of application, frequency of application, user activity, environmental factors, and arthropod species intended to repel (*42*).

The use of repellents continues to serve as the primary means of personal protection against biting arthropods (*90*). Repellents should be used in accordance with the label and are typically safe to apply to both skin and clothing. In fact, research shows that when avoidance of arthropod habitat is not an option, wearing protective clothing augmented with repellent applied to both clothing and skin is especially effective at preventing transmission of vector-borne disease (*17*, *90*, *91*). In most cases, the higher the concentration of active ingredient, the longer the duration of protection. However, application of products at concentrations > 50% does not appear to correlate with marked increased times of protection. Repellents should only be applied as needed or when the user begins to experience bites.

Self evaluation may also prove useful as repellents seem to provide varying levels of protection among individuals as observed in a 1999 study which reported that Deet provided nearly twice the repellent protection against *Anopheles stephensi* mosquitoes in men as compared to women (*92*). Repellent use is affected by industry, marketing, and word of mouth. In addition, repellents need to be user-friendly and have labels that are easy to read and understand. Consumers have stated that they want a product that is safe, effective, and cosmetic-friendly. Ultimately, the efficacy of a repellent as a frontline method to combat vector-borne diseases relies on a combination of factors. While science and industry will continue to research and produce new repellent compounds, acceptance and use of the repellent is solely dependent upon public compliance (*93*).

References

1. M. Debboun; S. P. Frances; D. Strickman *Insect Repellents: Principles, Methods, and Uses*, 1st ed.; CRC Press: Boca Raton, FL, 2007.
2. Fradin, M. S.; Day, J. F. Comparative efficacy of insect repellents against mosquito bites. *N. Engl. J. Med.* **2002**, *347*, 13–18.

3. Breman, J. G.; Alilio, M. S.; Mills, A. Conquering the intolerable burden of malaria. What's new, what's needed: A summary. *J. Trop. Med. Hyg.* **2004**, *71*, 1–15.
4. Goddard, J. *Physician's Guide to Arthropods of Medical Importance*, 3rd ed.; CRC Press: Boca Raton, FL, 2000.
5. Weaver, S. C.; Reisen, W. K. Present and future arboviral threats. *Antiviral Res.* **2009**.
6. Gage, K. L. Plague. In *Topley and Wilson's Microbiology and Microbial Infections*; Vol. 3, Bacterial Infections; Hausler, W. J., Jr., Sussman, M., Eds.; Hodder Arnold Publishers: London, 1998; pp 885–902.
7. Piesman, J.; Mather, T.; Dammin, G. J.; Telford, S. R.; Lastavica, C. C.; Spielman, A. Seasonal variation of transmission risk of Lyme disease and human babesiosis. *Am. J. Epidemiol.* **1987**, *126*, 1187–1189.
8. Goodman, J. L.; Nelson, C.; Vitale, B.; Madigan, J. E.; Dumler, J. S.; Kurtii, T. J.; Munderloh, U. G. Direct cultivation of the causative agent of human granulocytic ehrlichiosis. *N. Engl. J. Med.* **1996**, *334*, 209–215.
9. Stafford, K. C.; Massung, R. F.; Magnarelli, L. A.; Ijdo, J. W.; Anderson, J. F. Infection with agents of human granulocytic ehrlichiosis, Lyme disease, and babesiosis in wild white-footed mice (*Peromyscus leucopus*) in Connecticut. *J. Clin. Microbiol.* **1999**, *37*, 2887–2892.
10. Spach, D. H.; Liles, W. C.; Campbell, G. L.; Quick, R. E.; Anderson, D. E.; Fritsche, T. R. Tick-borne diseases in the United States. *New Engl. J. Med.* **1993**, *329*, 936–47.
11. Burgdorfer, W. Tick-borne diseases in the United States: Rocky Mountain spotted fever and Colorado tick fever. A review. *Acta Trop.* **1977**, *34*, 103–126.
12. Schulze, T. L.; Jordan, R. A.; Schulze, C. J.; Healy, S. P. Suppression of tick populations following annual habitat-targeted acaricide applications against fall populations of *Ixodes scapularis* (Acari: Ixodidae) in a residential landscape. *J. Am. Mosq. Control Assoc.* **2008**, *24*, 566–570.
13. Mount, G. A. A critical review of ultralow-volume aerosols of insecticide applied with vehicle-mounted generators for adult mosquito control. *J. Am. Mosq. Control Assoc.* **1998**, *14*, 305–334.
14. Gould, L. H.; Nelson, R. S.; Griffith, K. S.; Hayes, E. B.; Piesman, J.; Mead, P. S.; Carter, M. L. Knowledge, attitude, and behaviors regarding Lyme disease prevention among Connecticut residents, 1999-2004. *Vector-Borne Zoonotic Dis.* **2008**, *8*, 769–776.
15. Brown, M.; Hebert, A. A. Insect repellents: An overview. *J. Am. Acad. Dermatol.* **1997**, *36*, 243–249.
16. Katz, T. M.; Miller, J. H.; Miller, A. A. Insect repellents: Historical perspectives and new developments. *J. Am. Acad. Dermatol.* **2008**, *58*, 865–871.
17. Fradin, M. S. Mosquitoes and mosquito repellents: A clinician's guide. *Ann. Intern. Med.* **1998**, *128*, 931–940.
18. Bissinger, B. W.; Roe, R. M. Tick repellents: Past, present, and future. *Pestic. Biochem. Physiol.* **2010**, *96*, 63–79.

19. Maibach, H. L.; Akers, W. A.; Johnson, H. L.; et al. Insects: topical insect repellents. *Clin. Pharmocol. Ther.* **1974**, *16*, 970–973.
20. Garson, L. R.; Winnike, M. E. Relationship between insect repellency and chemical and physical parameters: A review. *J. Med. Entomol.* **1968**, *5*, 339–352.
21. Paluch, G. L.; Bartholomay, L.; Coats, J. Mosquito repellents: Chemical structure and olfaction. *Pest Manage. Sci.* **2010**, *66*, 925–935.
22. Bissinger, B. W.; Apperson, C. S.; Sonenshine, D. E.; Watson, D. W.; Roe, R. M. Efficacy of the new repellent BioUD® against three species of ixodid ticks. *Exp. Appl. Acarol.* **2009**, *48*, 239–250.
23. Centers for Disease Control and Prevention. Protection against Mosquitoes and Other Arthropods. httb//www.cdc.gov (accessed January 4, 2010).
24. Bock, G. R., Cardew, G., Eds.; *Olfaction in Mosquito-Host Interactions*; John Wiley: New York, 1996.
25. Khan, A. A. Mosquito Attractants and Repellents. In *Chemical Control of Insect Behavior*; Shorey, H. H., McKelvey, J. J., Eds.; John Wiley: New York, 1977, pp 305−325.
26. Schreck, C. E.; Kline, D. L.; Carlson, D. A. Mosquito attraction to substances from skin of different humans. *J. Am. Mosq. Control Assoc.* **1990**, *6*, 406–410.
27. Carroll, J. F.; Klun, J. A.; Debboun, M. Repellency of deet and SS220 applied to skin involves olfactory sensing by two species of ticks. *Med. Vet. Entomol.* **2005**, *19*, 101–106.
28. Pretorius, A. M.; Jensenius, M.; Clarke, F.; Ringertz, S. H. Repellent efficacy of deet and KBR 3023 against *Amblyomma hebreum* (Acari: Ixodidae). *J. Med. Entomol.* **2003**, *40*, 245–248.
29. Dautel, H. Test systems for tick repellents. *Int. J. Med. Microbiol.* **2004**, *293*, 182–188.
30. Peterson, C.; Coats, J. Insect repellents: Past, present and future. *Pestic. Outlook* **2001**, *12*, 154–158.
31. Moore, S. J.; Debboun, M. History of Insect Repellents. In *Insect Repellents: Principles, Methods, and Uses*; Debboun, M., Frances, S. P., Strickman, D., Eds.; CRC Press: Boca Raton, FL, 2007; pp 3−29.
32. Vernede, R.; van Meer, M. M.; Alpers, M. P. Smoke as a form of personal protection against mosquitoes: A field study in Papua New Guinea. *Southeast Asian J. Trop. Med. Public Health* **1994**, *25*, 771–775.
33. Schreck, C. E.; McGovern, T. P. Repellents and other personal protection strategies against *Aedes albopictus*. *J. Am. Mosq. Control Assoc.* **1989**, *5*, 247–250.
34. Barnard, D. R.; Bernier, U. R.; Posey, K. H.; Xue, R. D. Repellency of IR3535, KBR3023, para-menthane-3,8,-diol, and deet to black salt marsh mosquitoes (Diptera: Culicidae) in the Everglades National Park. *J. Med. Entomol.* **2002**, *39*, 895–899.
35. Barnard, D. R.; Xue, R. D. Laboratory evaluation of mosquito repellents against *Aedes albopictus, Culex nigripalpus*, and *Ochlerotatus triseriatus* (Diptera: Culicidae). *J. Med. Entomol.* **2004**, *41*, 726–730.

36. Constantini, C.; Badolo, A.; Ilboudo-Sanogo, E. Field evaluation of the efficacy and persistence of insect repellents containing deet, IR3535, and KBR 3023 against *Anopheles gambiae* complex and other Afrotropical vector mosquitoes. *Trans. R. Soc. Trop. Med. Hyg.* **2004**, *98*, 644–652.
37. Carroll, J. F.; Klun, J. A.; Debboun, M. Repellency of deet and SS220 applied to skin involves olfactory sensing by two species of ticks. *Med. Vet. Entomol.* **2005**, *19*, 101–106.
38. Hanifan, A. L.; Ismail, S. H.; Ming, H. T. Laboratory evaluation of four commercial repellents against larval *Leptotrombidium deliense* (Acari: Trombiculidae). *Southeast Asian J. Trop. Med. Public Health* **2010**, *41*, 1082–1087.
39. Klun, J. A.; Khrimian, A.; Rowton, E.; Kramer, M.; Debboun, M. Biting deterrent activity of a deet analog, two DEPA analogs, and SS220 applied topically to human volunteers compared with deet against three species of blood-feeding flies. *J. Med. Entomol.* **2006**, *43*, 1248–1251.
40. EPA: Insect Repellents: Use and Effectiveness. http://cfpub.epa.gov/oppref/insect/ (accessed January 2011).
41. Buescher, M. D.; Rutlegde, L. C.; Wirtz, R. A.; Nelson, J. H. The dose persistence relationship of deet against *Aedes aegypti. Mosq. News* **1983**, *43*, 364–366.
42. Goodyer, L. I.; Croft, A. M.; Frances, S. P.; Hill, N.; Moore, S. J.; Onyango, S. P.; Debboun, M Expert review on the evidence base for arthropod bite avoidance. *J. Trav. Med.* **2010**, *17*, 182–192.
43. Casida, J. E.; Quistad, G. B. *Pyrethrum Flowers: Production, Chemistry, Toxicology, and Uses*; Oxford University Press: Oxford, 1995.
44. Yue, S. J. *The Toxicology and Biochemistry of Insecticides*; Taylor & Francis Group, CRC Press: Boca Raton, FL, 2008.
45. Schreck, C. E.; Posey, K.; Smith, D. Durability of permethrin as a potential clothing treatment to protect against blood-feeding arthropods. *J. Econ. Entomol.* **1978**, *71*, 397–400.
46. Gupta, R. K.; Rutledge, L. C.; Reifenrath, W. G.; Gutierrez, G. A.; Korte, D. W., Jr. Effects of weathering on fabrics treated with permethrin for protection against mosquitoes. *J. Am. Mosq. Control Assoc.* **1989**, *5*, 176–179.
47. Schreck, C. E.; Mount, G. A; Carlson, D. A. Wear and wash persistence of permethrin used as clothing treatment for personal protection against the lone star tick (Acari: Ixodidae). *J. Med. Entomol.* **1982**, *19*, 143–146.
48. Rozendaal, J. A. Impregnated mosquito nets and curtains for self-protection and malaria control. *Trop. Dis. Bull.* **1998**, *86*, R1–R41.
49. Picaridin: A new insect repellent. *J. Drugs Dermatol.* **2004**, Jan-Feb.
50. Consumer Reports Confirms Effectiveness of New Alternative to Deet. *www.consumerreports.org./cro/cu-press-room/pressroom/archive/2005/07/eng0507det.htm*.
51. Frances, S. P.; Cooper, R. D. Personal protection measures against mosquitoes: A brief history and current use of repellents by the Australian Defense Force. *ADF Health* **2002**, *3*, 58–63.
52. Frances, S. P.; Waterson, D. G. E.; Beebe, N. W.; Cooper, R. D. Field evaluation of repellent formulations containing deet and picaridin against

mosquitoes in Northern Territory, Australia. *J. Med. Entomol.* **2004**, *41*, 414–417.

53. Insect repellents: Which keep bugs at bay? *Consumer Reports* **2006**, *71*, 6.
54. Stafford, K. C., III. Tick Bite Prevention, 2006. The Connecticut Agricultural Experiment Station. http/www.dph.state.ct.us/BCH/infectiousdise/tickborne/tick/htm#DEET%20and%20ticks (accessed March 7, 2010).
55. Yap, H. H.; Jahangir, A.; Chong, S. C.; et al. Field efficacy of a new repellent KBR 3023 against *Aedes albopictus* (Skuse) and *Culex quinquefasciatus* (Say) in a tropical environment. *J. Vector Ecol.* **1998**, *23*, 62–63.
56. Carroll, J. F.; Benante, J. P.; Klun, J. A. Twelve-hour duration testing of cream formulations of three repellents against *Amblyomma americanum*. *Med. Vet. Entomol.* **2008**, *22*, 144–151.
57. Moore, S. J.; Lenglet, A.; Hill, N. Plant-Based Insect Repellents. In *Insect Repellents: Priniciples, Methods, and Uses*; Debboun, M., Frances, S. P., Strickman, D., Eds.; CRC Press: Boca Raton, FL, 2007; pp 245−259.
58. EPA: Citronella Factsheet. http://www.epa.gov/opp00001/biopesticides/ingredients/factsheets/factsheet_021901.htm (accessed March 2011).
59. Kongkaew, C.; Sakunrag, I.; Chaiyakunapruk, N.; Tawatsin, A. Effectiveness of citronella preparations in preventing mosquito bites: Systematic review of controlled laboratory experimental studies. *Trop. Med. Int. Health* **2011**, *16*, 802–810.
60. Tawatsin, A.; Wratten, S.; Scott, R.; Thavara, U.; Tachadamrongsin, Y. Repellency of volatile oils from plants against three mosquito vectors. *J. Vector Ecol.* **2001**, *26*, 76–82.
61. Suwonkerd, W.; Tantrarongroj, K. Efficacy of essential oil against mosquito biting. *Commun. Dis. J.* **1994**, *20*, 4–11.
62. Amer, A.; Mehlhorn, H. Repellency effect of forty one essential oils against *Aedes, Anopheles* and *Culex* mosquitoes. *Parasitol. Res.* **2006**, *99*, 478–490.
63. Osmani, Z.; Anees, I.; Naidu, M. Insect repellent creams form essential oils. *Pesticides* **1972**, *6*, 9–21.
64. EPA: Regulating Biopesticides. http/www.epa.gov/oppbppd1/biopesticide/index.htm (accessed March 2011).
65. Puccetti, G. IR3535 (Ethyl Butylacetylaminoproprionate). In *Insect Repellents: Principles, Methods, and Uses*; Debboun, M. Frances, S. P., Strickman, D., Eds.; CRC Press: Boca Raton, FL, 2006; pp 353−360.
66. Naucke, T. J.; Lorentz, S.; Grünwald, H. W. Laboratory testing of the insect repellents IR3535 and deet against *Phlebotomus mascittii* and *P. duboscqi* (Diptera: Psychodidae). *Int. J. Med. Microbiol.* **2006**, *296*, 230–232.
67. Carroll, S. P. Topical Insect Repellents and Factors That Affect Their Performance. In *Insect Repellents: Principles, Methods, and Uses*; Debboun, M., Frances, S. P., Strickman, D., Eds.; CRC Press: Boca Raton, FL, 2006; pp 245−259.
68. Carroll, S. P. Prolonged efficacy of IR3535 repellents against mosquitoes and blacklegged ticks in North America. *J. Med. Entomol.* **2008**, *45*, 706–714.
69. Lindsay, R. L.; Heal, J. D.; Surgeoner, G. A. *Comparative Evaluation of the Efficacy of Bite Blocker, Off! Skintastic, and Avon Skin-So-Soft To Protect against Aedes Species Mosquitoes in Ontario*; Department of Environmental

Biology, University of Guelph, Sponsored by Chemfree Environment, Inc.: Guelph, Ontario, 1996.

70. Barnard, D. R. Repellency of essential oils to mosquitoes (Diptera: Culicidae). *J. Med. Entomol.* **1999**, *36*, 625–629.
71. Trongtokit, Y.; Rongsriyam, R.; Komalamisra, N.; Apiwathnasorn, C. Comparative repellency of 38 essential oils against mosquito bites. *Phytother. Res.* **2005**, *19*, 303–309.
72. Carroll, S. P.; Loye, J. E. PMD: A registered botanical mosquito repellent with deet-like efficacy. *J. Am. Mosq. Control Assoc.* **2006**, *22*, 507–513.
73. Curtis, C. F. *Control of Disease Vectors in the Community*; Wolfe Publishing: London, 1991.
74. Curtis, C. F. Personal protection methods against vectors of disease. *Rev. Med. Vet. Entomol.* **1992**, *80*, 543–553.
75. Schreck, C. E.; Leonhardt, B. A. Efficacy assessment of quwenling: A mosquito repellent from China. *J. Am. Mosq. Control Assoc.* **1991**, *7*, 433–436.
76. Collins, D. A.; Brady, J. N.; Curtis, C. F. Assessment of the efficacy of quwenling as a mosquito repellent. *Phytother. Res.* **1993**, *7*, 17–20.
77. Trigg, J. K. Evaluation of a eucalyptus-based repellent against *Anopheles* spp. in Tanzania. *J. Am. Mosq. Control Assoc.* **1996**, *12*, 243–246.
78. Farrar, R. R.; Kennedy, G. G. 2-Undecanone, a constituent of the glandular trichomes of *Lycopersicon hirsutum f. glabratum*: Effects on *Heliothis zea* and *Manduca sexta* growth and survival. *Entomol. Exp. Appl.* **1987**, *43*, 17–23.
79. Bissinger, B. W.; Zhu, J.; Apperson, C. S.; Sonenshine, D. E.; Watson, D. W.; Roe, R. M. Comparative efficacy of BioUD to other commercially available arthropod repellents against the ticks *Amblyomma americanum* and *Dermacentor variabilis* on cotton cloth. *Am. J. Trop. Med. Hyg.* **2009**, *81*, 685–690.
80. Bissinger, B. W.; Apperson, C. S.; Sonenshine, D. E.; Watson, D. W.; Roe, R. M. Efficacy of the new repellent BioUD® against three species of ixodid ticks. *Exp. Appl. Acarol.* **2009**, *48*, 239–250.
81. Witting-Bissinger, B. E.; Stumpf, C. F.; Donohue, K. V.; Apperson, C. S.; Roe, R. M. Novel arthropod repellent, BioUD, is an efficacious alternative to deet. *J. Med. Entomol.* **2008**, *45*, 891–898.
82. Panella, N. A.; Dolan, M. C.; Karchesy, J. J.; Xiong, Y.; Peralta-Cruz, J.; Khasawneh, M.; Montenieri, J. A.; Maupin, G. O. Use of novel compounds for pest control: Insecticidal and acaricidal activity of essential oil components from heartwood of Alaska yellow cedar. *J. Med. Entomol.* **2005**, *42*, 352–358.
83. Dolan, M. C.; Jordan, R. A.; Schulze, T. L.; Schulze, C. J.; Manning, M. C.; Ruffalo, D. R.; Schmidt, J. P.; Piesman, J.; Karchesy, J. J. Ability of two natural products, nootkatone and carvacrol, to suppress *Ixodes scapularis* and *Amblyomma americanum* (Acari: Ixodidae) in a Lyme disease endemic area of New Jersey. *J. Econ. Entomol.* **2009**, *102*, 2316–2324.
84. Dietrich, G.; Dolan, M. C.; Peralta-Cruz, J.; Schmidt, J.; Piesman, J.; Eisen, R. J.; Karchesy, J. J. Repellent activity of fractioned compounds from

Chamaecyparis nootkatensis essential oil against nymphal *Ixodes scapularis* (Acari: Ixodidae). *J. Med. Entomol.* **2006**, *43*, 957–961.

85. Schulze, T. L.; Jordan, R. A.; Dolan, M. C. Experimental use of two standard tick collection methods to evaluate the relative effectiveness of several plant-derived and synthetic repellents against *Ixodes scapularis* and *Ambloymma americanum* (Acari: Ixodidae). *J. Econ. Entomol.* **2011**, in press.
86. Jordan, R. A.; Schulze, T. L.; Dolan, M. C. Efficacy of plant-derived and synthetic compounds on clothing as repellents against *Ixodes scapularis* and *Amblyomma americanum* (Acari: Ixodidae). *J. Med. Entomol.* **2011**, in press.
87. Tong, F.; Gross, A. D.; Dolan, M. C.; Coats, J. R. Effects of monoterpenoid insecticide/acaricide carvacrol, on [^{14}C]- nicotine binding to the *Musca domestica* (house fly) nicotine acetylcholine receptor. *Insect Biochem. Mol. Biol.* **2011**, in review.
88. Sudakin, D. L.; Trevathan, W. R. DEET: A review and update of safety and risk in the general population. *J. Toxicol. Clin. Toxicol.* **2003**, *41*, 831–839.
89. Aquino, M.; Fyfe, M.; MacDougal, L.; Remple, V. West Nile virus in British Columbia. *Emerging Infect. Dis.* **2004**, *10*, 1499–1501.
90. Piesman, J.; Eisen, L. Prevention of tick-borne diseases. *Annu. Rev. Entomol.* **2008**, *53*, 323–343.
91. Vasquez, M.; Muehlenbein, C.; Cartter, M.; Hayes, E. B.; Ertel, S.; Shapiro, E. D. Effectiveness of personal protection measures to prevent Lyme disease. *Emerging Infect. Dis.* **2008**, *14*, 210–216.
92. Golenda, C. F.; Solberg, V. B.; Burge, R.; Gambel, J. M.; Wirtz, R. A. Gender-related efficacy difference to an extended duration formulation of topical N,N-Diethyl-*m*-Toluamide (DEET). *Am. J. Trop. Med. Hyg.* **1999**, *60*, 654–657.
93. Strickman, D.; Frances, S. P.; Debboun, M. Epilogue: Prospects for the Future. In *Insect Repellents: Principles, Methods, and Uses*; Debboun, M., Frances, S. P., Strickman, D., Eds.; CRC Press: Boca Raton, FL, 2006; pp 425−427.

Chapter 2

Development of Novel Repellents Using Structure–Activity Modeling of Compounds in the USDA Archival Database

Ulrich R. Bernier* and Maia Tsikolia

Center for Medical, Agricultural, and Veterinary Entomology, Agricultural Research Service, United States Department of Agriculture, 1600 SW 23rd Drive, Gainesville, Florida, 32608
***E-mail: uli.bernier@ars.usda.gov**

The United States Department of Agriculture (USDA) has developed repellents and insecticides for the U.S. military since 1942. Repellency and toxicity data for over 30,000 compounds are contained within the USDA archive. Repellency data from subsets of similarly structured compounds were used to develop artificial neural network (ANN) models to predict new compounds for testing. Compounds were then synthesized and evaluated for their repellency against *Aedes aegypti* mosquitoes. Rellency data, *i.e.*, complete protection time (CPT) were used to develop Quantitative Structure Activity Relationship (QSAR) models to predict repellency. Successful prediction of novel acylpiperidine structures by ANN models resulted in the discovery of compounds that provided protection more than three times longer than DEET. The acylpiperidine QSAR models employed 4 descriptors to describe the relationship between structure and repellent duration. The ANN model of the carboxamides did not predict compound structures with exceptional CPTs as accurately; however, several carboxamide candidates did perform as good as or better than DEET.

History of the USDA Evaluation Program

In March 1942, funds from the National Emergency Council, Office of Scientific Research and Development (NEC-OSRD) were made available to the Bureau of Entomology and Plant Quarantine, Division of Insects Affecting Man and Animals, of the United States Department of Agriculture (USDA)-Agricultural Research Administration to expand the small field laboratory in Orlando, FL. Willard V. King was appointed to oversee the development of the Orlando laboratory. By the time the laboratory was fully operational, W.E. Dove had assumed the role of director, and he was followed by Ed Knipling in July, 1942 (*1*). The mission of the laboratory was to discover new chemicals and methods for the control of medically-important arthropod pests of the U.S. Armed Forces (*1–3*). Most of the early submissions received for screening by the Orlando laboratory for screening consisted of known commercially available insecticides and repellents either submitted by commercial entities, the Bureau of Entomology Insecticide Investigations, or by other agencies of the US Government as part of the OSRD. The program was expanded in June, 1944 to include Columbia, Harvard, Ohio State, and Stanford Universities, along with the Universities of Illinois, Maryland, Minnesota, and Wisconsin to provide candidate compounds for evaluation at the Orlando laboratory (*2*). On November 1, 1945, the source of funding for the program was changed to the U.S. Army Office of the Surgeon General, and in the late 1950s this funding line was transferred to the Insects Affecting Man and Animals Branch of the USDA-Entomology Research Division with the expectation that the program would continue the development of control methods to protect US service personnel from arthropod attack. The Orlando laboratory has changed names and locations throughout the years. In 1951, it was renamed the "Insects Affecting Man and Animals Research Laboratory (IAMARL), along with the formation of the "Mosquito Research Unit." The laboratory was moved from Orlando, FL, to Gainesville, FL, in 1962. The unit conducting mosquito research was renamed the "Mosquito and Fly Research Unit" in 1988. The laboratory was renamed the Medical and Veterinary Entomology Research Laboratory (MAVERL) in 1990, and finally the Center for Medical, Agricultural, and Veterinary Entomology (CMAVE) in 1996.

Early History of the USDA Repellent and Insecticide Program and Archive

On March 11, 1942, the Orlando laboratory of the Bureau of Entomology recorded its first chemical submission and using the code O-1 (Orlando-1). The sample consisted of six 10-oz jars of "Pyrinate" from McKesson & Robbins. Most of the submissions received over the next two months were pyrethrin mixtures and chlorinated hydrocarbons. The well known insecticide 1,1,1-trichloro-2,2-bis(4-chlorophenyl)ethane (DDT) was submitted as the active ingredient in two products named "Gesarol." The two products, O-1151 (dust) and O-1152 (spray), were logged into one of the archival record books on November 16, 1942, with information that the sample was received from the New York division of the J.R. Geigy Company (*3*).

Some of the best repellent active ingredients during this time period were Indalone (O-9), tested on March 27, dimethyl phthalate (O-262), tested on May 8, and Rutgers 612 (O-375), tested on June 15, 1942. These three repellents were mentioned by Ed Knipling as the best from the initial screening phase and they were recommended for U.S. Military use since they provided about 2 h protection time (*3*). However, military personnel still needed a repellent that would last for 10 hours. This need was met about a decade later when the most successful mosquito repellent to date, N,N-diethyl-3-methylbenzamide (DEET; Figure 1) was recorded as O-20218 on February 5, 1952. It had been sent from S.A.Hall, a chemist with the division of insecticide investigations, Bureau of Entomology and Plant Quarantine, at Beltsville, MD, to the Orlando laboratory and received by William C.McDuffie, assistant leader of the Insects Affecting Man and Animals Section of Entomology Research Branch. DEET was first screened as a clothing treatment and found to be a superior candidate (*4*). This led to its selection for field trials conducted in Panama in early 1953 (*5*). A second submission for DEET was recorded as O-22542 on December 17, 1953 and the following is written in the notebook:

> *"Reaction product of mixed toluic acid isomers (containing approx. 70% m-toluic acid and 30% p-toluic acid) and diethylamine. Insecticide Investigations, Memo S.A. Hall to W.C.M. December 14, 1953, 50g."*

Under the USDA archival record system, the final compound submitted by the Beltsville laboratory was recorded in the notebook as AI3-55208 (formerly Orlando numbers, now AI3- numbers for "Agricultural Insecticide 3-"). It was sent by Al DeMilo of the Beltsville Laboratory on May 22, 1997, and tested by Don Barnard and his group in Gainesville on September 8, 1997. In actuality, sublots of formerly tested compounds continued to be received from Beltsville. The final recorded entry is for AI3-37220-Gf on May 12, 1998 and this compound will be discussed further later in this chapter.

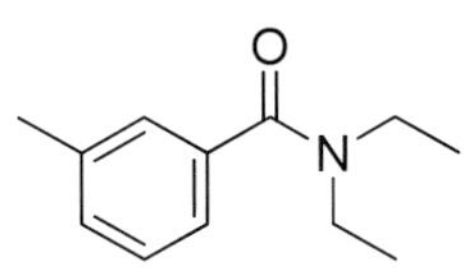

(O-20218)

Figure 1. Orlando number and structure of N,N-diethyl-3-methylbenzamide (DEET).

Recent Research of the USDA Insecticide and Repellent Program

Research by the USDA for the U.S. Military continued with lower intensity through the 1990s. A significant stimulus to reinvigorate the association between the USDA-Agricultural Research Service (ARS) and the U.S. Military was made possible in 2004 by a new Department of Defense funding line named the "Deployed War-Fighter Protection Program" (DWFP). The emphasis of the research program is on the development of novel or improved pesticide chemicals and formulations, application technologies, and personal protection. It is within the realm of personal protection that the research and development of novel topical repellents is being conducted. There are several sources that provide repellents as part of this renewed collaborative effort. Among these are the USDA-ARS laboratories in Beltsville, MD, (Invasive Insect Biocontrol and Behavior Laboratory), Gainesville, FL (CMAVE) and Oxford, MS (Natural Products Utilization Research Unit), researchers at the University of Mississippi, and in Australia, French Polynesia, Germany, Isreal, Malaysia, Samoa, Saudi Arabia, and Turkey. While one objective of the laboratory in Gainesville, FL, is to provide insecticide and repellent evaluation for the DWFP, researchers at this laboratory are also devoted to the development of new repellents using the data contained in the historical archive. Through the use of modern methods of structure–activity modeling, the goals are: a) to understand better how chemical structure relates to repellency by developing accurate models and b) to develop improved repellents as an outcome of these models. This work involves collaboration with chemists at the University of Florida and the results of this effort are the subject of this chapter.

Structure–Activity and Computer Modeling

The examination of molecular structures and modeling can be traced back to the early 1900s (*6*). Prior to the development of computers, the examination of a set of chemicals and attempts to relate their activity to the structures was almost entirely dependent on the skill of the synthetic chemist to devise and synthesize structurally-similar compounds once a lead compound was identified. Upon examination of the archive, it is evident that the discovery of the repellent DEET was due to a process where related structures had been evaluated and found to be repellents. In the spring of 1952, the compounds *N,N*-diethylbenzamide (O-1197-d) and *o*-chloro-*N,N*-diethylbenzamide (O-17586-b) (Figure 2) were tested on skin against *Aedes aegypti*, with the latter compound protecting about 10% longer than the former (*4*). The compound *N,N*-diethylbenzamide had been received from the USDA Beltsville Insecticide Division and logged in originally on November 23, 1942, during the first year of the program. The O-17586-b compound noted above was received from Geigy Co.; however, this compound was originally received from the Insecticide Division and tested on March 29, 1946 (originally recorded as O-11147). Samuel Gertler applied for a patent covering the *N,N*-diethylbenzamides as repellents on September 4, 1944 and the patent was granted on Oct. 1, 1946 (*7*). Unfortunately, these compounds mentioned above produced skin irritation, so further repellent studies with them were abandoned. The continued efforts to produce structurally-similar substituted *N,N*-diethylbenzamides by the Beltsville chemists led to the discovery of DEET as one of the best repellents as noted in McCabe et al. (*8*).

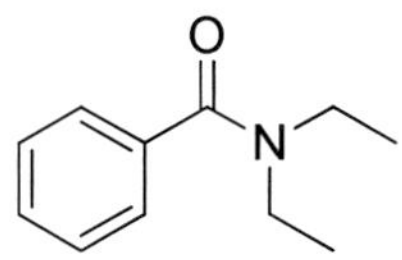

(O-1197-d)
N,N-diethylbenzamide

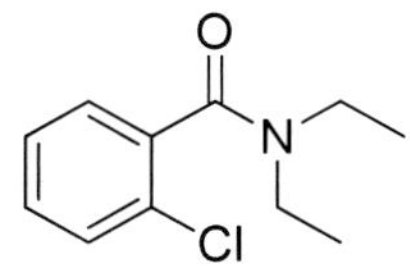

(O-17586-b)
o-chloro-*N,N*-diethylbenzamide

Figure 2. Repellents tested prior to the discovery of DEET.

Uses of Computer Modeling of Chemical Structures

Extensive use of computers in modeling began in the 1950s (*9*). The specifics of linear modeling of biological properties, specifically Quantitative Structure–Activity Relationships (QSAR), can be traced back to the work of Corwin Hansch and colleagues in the early 1960s (*10*). As noted by Hansch, contrary to the belief that the history and success of QSAR lies in the pharmaceutical domain, the earliest applications and successes involved the modeling of pesticides.

Application of Modeling Methods to Repellent Discovery

Structure–activity modeling has also been applied to repellent discovery, with perhaps one of the greater successes being the discovery of 2-(2-hydroxyethyl)-1-piperidinecarboxylic acid 1-methylpropyl ester, more commonly known by the names Picaridin, Icaridin, KBR 3023, or Bayrepel ® (Figure 3). This compound was discovered through structure–activity work in the 1980s (*11*) and tested as AI3-65545 on October 31, 1993.

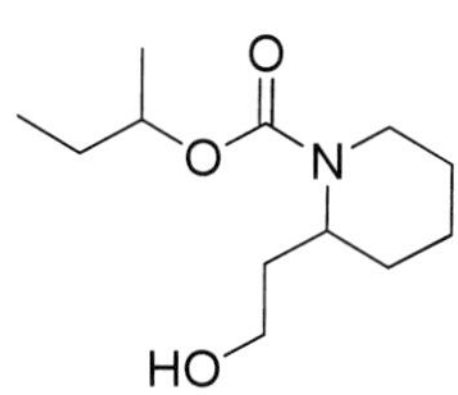

(AI3-65545)

Figure 3. Structure of Picaridin (KBR 3023)

Researchers have used 3D-QSAR of DEET and related analogues to construct pharmacophores to better understand the structural basis that leads to repellency by these amide compounds (*12–14*). Their model was constructed primarily from the protection time data of Suryanarayana et al. (*15*). Ma et al. (*12*) demonstrated that one could predict repellent duration based on compound structure, and specifically that the amide group and attached substituents played a significant role in the experimentally determined repellent efficacy. Using the same data set, Katritzky et al. (*16*) applied Codessa Pro software (*17*) to develop a QSAR model for the prediction of complete protection time (CPT) from descriptors related to the structural and electronic properties of the DEET analogues. This work was the foundation for current projects that involve examination of repellency and toxicity data for subsets of compounds within the USDA Archive.

However, there is a weakness in the way that repellency data are recorded in the USDA archive and this impacts the development of structure–activity models. Instead of being reported in days or time of CPT, the repellent protection times were converted to a 5 class system based on CPT as detailed in Table I. The groupings are not only non-linear but tend to equate all superior repellents (class 5) as identical to one another when in fact there can be significant differences in numbers of days that compounds are repellent.

Table I. Five class system of repellents based on complete protection time (CPT) from treated cloth and stockings. SOURCE: Reproduced from reference (*18*). Copyright 2010 Entomological Society of America

Class	*Minimum Day*	*Maximum Day*
1	0	1
2	1	5
3	5	10
4	10	21
5	21	–

Fortunately, artificial neural networks (ANNs) can overcome these limitations and can be used to develop models for these types of data. Some of the earliest work with neural networks was that of McCulloch and Pitts in 1943 (*19*). They can be used for evaluation of non-linear data for the development of a predictive model. Thus, a non-linear data set, such as the class system of CPT data in the USDA archive, can be used to develop a model and predict compound activities based on the compound structures and associated repellent activities that were incorporated into the neural network.

Three-layer neural networks with different architectures were applied to the two data sets discussed in this chapter, *i.e*,. acylpiperidines and carboxamides.

Development of the ANN model was the first step used to predict new repellents. This was accomplished by selecting a set of similarly structured compounds from the USDA archive, then randomly dividing the compounds into a training set and a validation set. The training set contained approximately 75% of the compounds used to develop the model. The remaining compounds were then used as the validation set to verify the accuracy of the model. If there was good correlation between predicted values (classes in the case of repellents) and the experimentally determined class, then the ANN was used to predict classes for compound structures that are input into the model. Some predicted structures were synthesized, and then evaluated for repellent efficacy by measurement of CPT, and in the case of the carboxamides, by both CPT and the minimum effective dosage (MED), which is the concentration required to produce repellency. Rather than converting these data to classes as had been done historically, the actual number of days of protection, or the threshold concentration of protection was used in efforts to develop QSAR models.

Measurement of Repellent Efficacy

Screening for Repellency of Compounds with Unknown Toxicology

In screening studies, approximately 500 colony-reared female *Ae. aegypti* (Orlando strain, 1952), aged 5-10 days and maintained on 10% sugar solution, were used per cage (approximately 46 cm x 36 cm x 36 cm ≈ 59,000 cm^3). Since stock cages of mosquitoes contain both males and females, a drawbox was used to preselect females that responded to human odors with the appropriate host-seeking behavior (*20*).

Because the experimental compounds screened in these studies have unknown toxicology, they should not be applied directly to the skin. However, muslin cloth can be treated with the candidate as a means to test the compound without topical application (*21*). Compounds are placed in separate vials and dissolved into a solvent that evaporates rapidly, *e.g.* acetone. A 5 cm x 10 cm segment of muslin cloth is then added to the vial containing the compound in solution. The cloth is removed and dried until the solvent evaporates. When ready to be tested, a volunteer can affix the treated cloth to cover a 32 cm^2 opening on a specially designed vinyl sleeve (Figure 4). The hand of the volunteer is gloved to protect from bites, and the only accessible area for mosquitoes to bite is through the opening in the sleeve. The cloth does not come in direct contact with the skin because of a stocking worn underneath the sleeve to provide a small barrier between the cloth and skin. The use of skin emanations is needed to attract mosquitoes to the opening in the vinyl sleeve. However, just as with other laboratory-based screening methods, the performance of a compound on cloth only partially reflects what the performance would be like if applied directly on skin.

Since these studies involved human volunteers, all participants were required to provide informed consent to participate. All data were collected in accordance with the approved University of Florida Institutional Review Board (UF IRB) Project entitled, "Laboratory Evaluation of Repellents for Personal Protection from Mosquitoes and Biting Flies" (Project # 636-05).

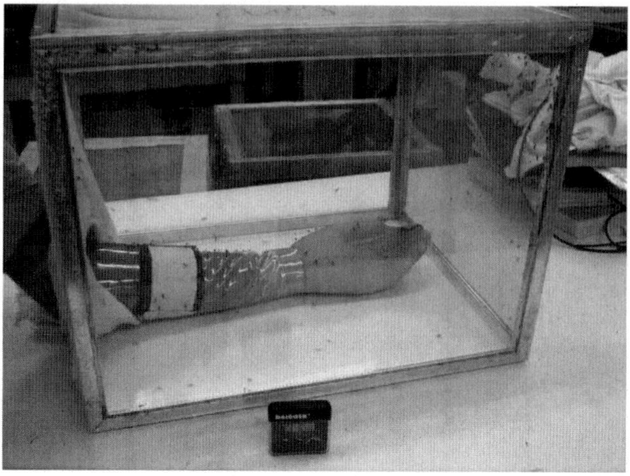

Figure 4. Photo Credit: Greg Allen, USDA-ARS. Reproduced from reference (18). Copyright 2010 Entomological Society of America. (see color insert)

Duration of Repellent Efficacy

The repellency duration is measured by the complete protection time (CPT), which is the amount of time in days that a compound will fully protect the wearer from bites of a test population of mosquitoes or other biting arthropods. In the case of mosquitoes, the end point is normally measured as the "time to first bite," however, quite often a second bite is used to provide the "time to the first confirmed bite" (*22*). There are concerns about significant errors resulting from measurement of a single bite as an end point despite the CPT being a useful and understandable metric to compare repellent efficacies. Therefore, the end point is normally selected to be a threshold number of bites. In the experiments described here, the failure threshold was predetermined to be the point at which 1% of mosquitoes had bit through the cloth (5 bites out of the 500 mosquitoes in the cage) during the 1 min test period. The CPTs were determined at 25 $\mu mol/cm^2$ and 2.5 $\mu mol/cm^2$ concentrations (*18, 23*). These concentrations were selected to bracket the amount of DEET that is typically applied directly to skin in repellency assays.

Threshold Concentration for Repellency

The threshold amount of a repellent needed to prevent bites is estimated by measuring the minimum effective dosage (MED) of the repellent (*18*)(*21*). A range of concentrations on cloth was used in these experiments starting with a high concentration of 25 $\mu mol/cm^2$. Serial dilutions were made from 25 $\mu mol/cm^2$ down to 3.125 $\mu mol/cm^2$ using the higher concentration solution, and from 2.5 $\mu mol/cm^2$ down to 0.020 $\mu mol/cm^2$ using the lower concentration solution. Similar to tests for CPT, the arm with treated cloth was inserted into the mosquito cage and tested for 1 min. If < 5 bites are received (1% out of 500), then the compound is considered repellent at that concentration.

Acylpiperidine Repellents

Acylpiperidine repellents have been studied for decades. Picaridin (Figure 3), the active ingredient in a number of commercial products, belongs to this class of compounds. Two of the more efficacious experimental repellents discovered by the USDA Beltsville laboratory are also in this class: 1-(cyclohex-3-en-1-ylcarbonyl)piperidine (AI3-35765) and 1-(cyclohex-3-en-1-ylcarbonyl)-2-methylpiperidine (AI3-37220) (Figure 5).

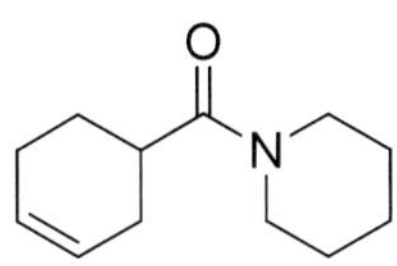

(AI3-35765)
cyclohex-3-en-1-ylcarbonyl)piperidine

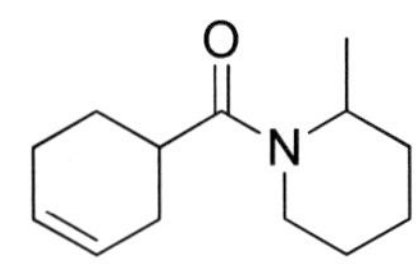

(AI3-37220)
1-(cyclohex-3-en-1-ylcarbonyl)-2-methylpiperidine

Figure 5. Piperidine repellents developed the USDA in the 1970s.

The AI3-35765 compound was tested on April 17, 1973, having been sent from the Organic Chemistry Synthesis Laboratory (OCSL) of the USDA Beltsville laboratory. On April 26, 1977, AI3-37220 was tested after it was synthesized by Terry McGovern of the OCSL (*24*). Later 3D-QSAR studies on Picaridin and 1-(cyclohex-3-ene-1-ylcarbonyl)-2-methylpiperidine (AI3-37220) using a hierarchical molecular overlay approach showed the importance of shape and molecular surface structure for effective repellent activity in the diastereoisomeric compounds of AI3-37220 (*25*). Calculations for the most active diastereoisomer (220SS) identified by Klun et al. (*26*) indicated a strong relationship between the structure and the biological potency.

Artificial Neural Network Modeling

The initial repellent model for the acylpiperidine data set was developed using 150 out of 200 selected acylpiperidines as the training set for the ANN. A full listing of the compounds, (coded by AI3- numbers), structural information, and notation of whether they were in the training or validation subsets can be found in the Supporting Information for Katritzky et al. (*23*). This set did not include AI3-35765 or AI3-37220 in the model, but did contain some compounds similar to those in structure (see Table II, *e.g.* **4a′-4d′** and others). The archival data used for the initial models in this study were accumulated from compounds submitted as early as 1942 and as late as 1994; the compound structures with AI3- numbers can be found in Table S1of the supplementary information of Katritzky et al. (*23*). Some of the modeled compounds were from acylpiperidines patented as insect repellents in 1981 (*27*).

The models for the acylpiperidines were developed with an 8-7-1 architecture, comprised of 8 initial descriptors as neurons for the input layer, followed by 7 neurons in a hidden layer, and the output of the predicted class as the final neuron. The input descriptors used to produce the best model were: 1) 3rd order Kier and Hall index, 2) molecular weight, 3) molecular surface area, 4) total molecular dipole moment, 5) total molecular electrostatic interaction, 6) total number of bonds in the molecule, 7) carbon atom surface area, and 8) nitrogen atom surface area. The resultant ANN model was able to predict the most efficacious repellents (class 4 and 5) with 71% accuracy (*23*).

With a satisfactory ANN model, structures can be devised and tested in the model to predict their repellent classes. This was performed with just over 2000 acylpiperidine structures. Some of these compounds had been tested previously, but many others were novel in the sense that they had not been evaluated previously as mosquito repellents. From 2000 predicted compounds, 34 were selected for synthesis: 23 were novel compounds, and 11 were chosen from those in the USDA archive. Selection of compounds that had been tested previously allowed for comparison and validation of the current repellent testing methodology with that used decades ago. The repellency data generated for this study were more precise and linear, *i.e.*, the repellency was measured in days of protection, rather than put into classes with non-linear distributions of protection time. Also, bioassays were conducted with stoichiometrically equivalent amounts of compounds, rather than comparison of gravimetrically equivalent amounts, as had been done historically. Generating data based on these changes was necessary for development of accurate QSAR models.

Synthesis

The selected 34 acylpiperidine mosquito repellent candidates **4a-q′** were synthesized according to the pathway of Figure 6 (*23*). Treatment of the carboxylic acids **1** with thionyl chloride and benzotriazole at 25 °C in methylene chloride in a 1:1:3 mole ratio produced 1-acylbenzotriazoles **2** (*23*). Reaction of 1-acylbenzotriazoles **2** with one equivalent of piperidines **3** in tetrahydrofuran, THF at 25 °C or in toluene under reflux resulted in formation of N-acylpiperidines **4a-q′**(Table II) in 71–100% yields using a procedure modified slightly from one used historically (*28*).

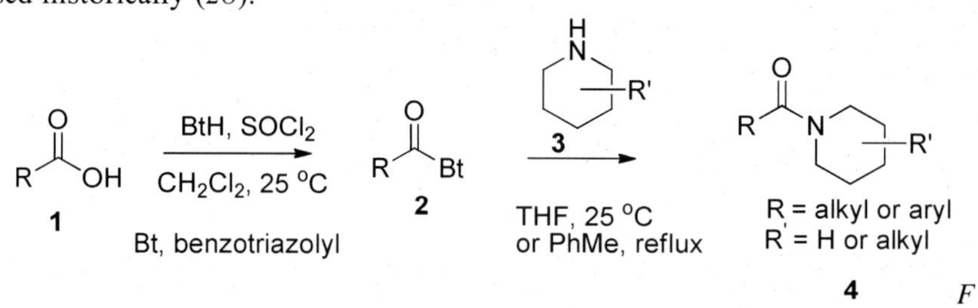

*Figure 6. Preparation of acylpiperidines **4**. Reproduced from reference (23). Copyright 2008 The National Academy of Sciences of the USA.*

Table II. Compounds used for the acylpiperidine repellent study

ID	*Name*	*Structure*
DEET	N,N-diethyl-3-methylbenzamide (O-20218)	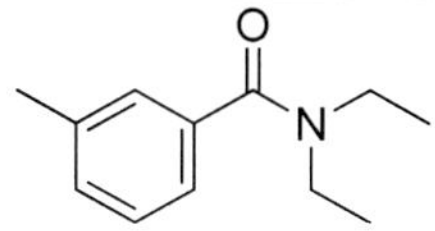
4a[a]	1-acetyl -2-methylpiperidine	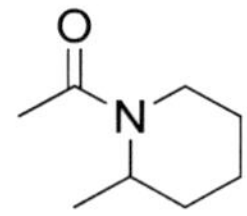
4b[a]	1-(1-oxopropyl)-piperidine	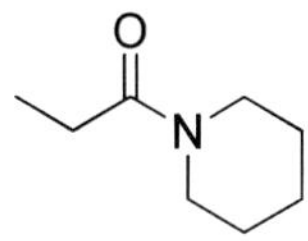
4c[a]	2-ethyl-1-(1-oxopropyl)-piperidine	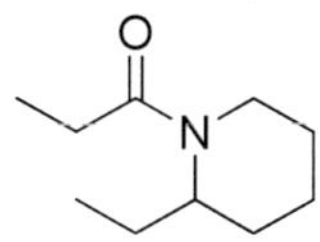
4d[a]	2-methyl-1-(1-oxoheptyl)-piperidine	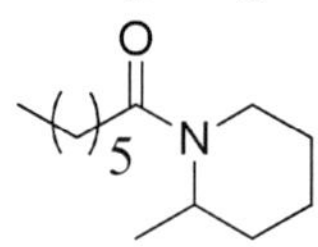
4e[a]	3-methyl-1-(1-oxoheptyl)piperidine	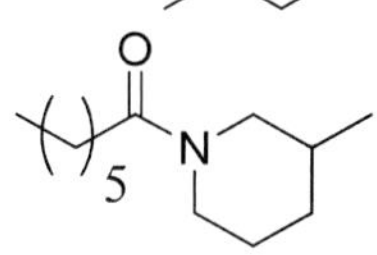
4f[a]	4-methyl-1-(1-oxooctyl)piperidine	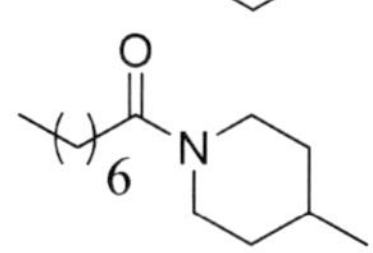
4g[a]	1-(1-oxooctyl)-4-(phenylmethyl)piperidine	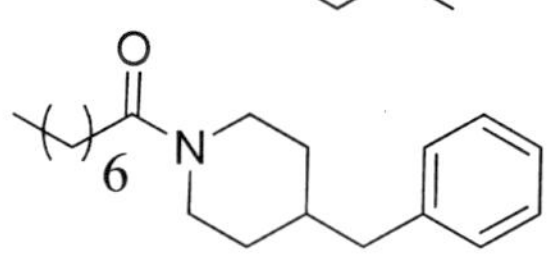
4h[a]	2-ethyl-1-(1-oxononyl)piperidine	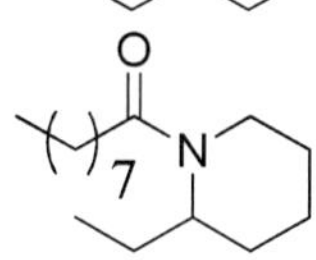
4i[a]	2-methyl-1-(1-oxodecyl)piperidine	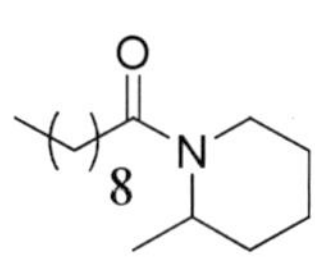
4j[a]	4-methyl-1-(1-oxodecyl)piperidine	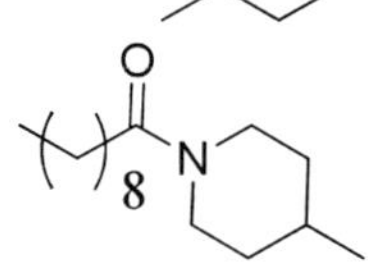

Continued on next page.

Table II. (Continued). Compounds used for the acylpiperidine repellent study

4k	1-(1-oxo-10-undecylenyl)piperidine AI3-39049	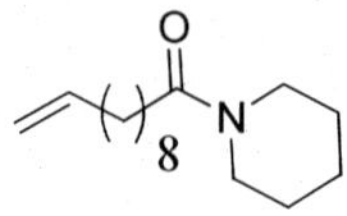
4l	2-ethyl-1-(1-oxo-10-undecylenyl)piperidine	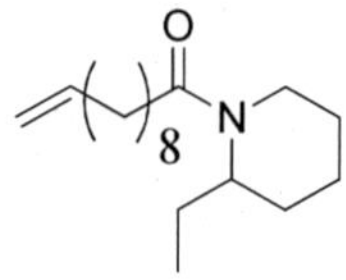
4m[a]	1-(1-oxo-10-undecylenyl)-4-(phenylmethyl)piperidine	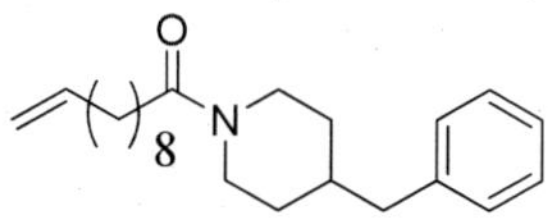
4n[a]	4-methyl-1-(1-oxo-10-undecylenyl)piperidine	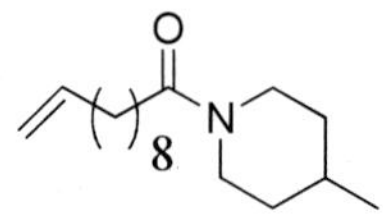
4o[a]	1-(1-oxoundecyl)piperidine	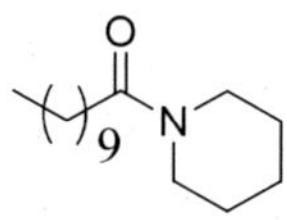
4p[a]	2-methyl-1-(1-oxododecyl)piperidine	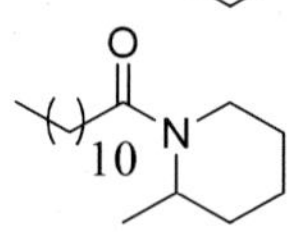
4q[a]	3-methyl-1-(1-oxododecanyl)piperidine	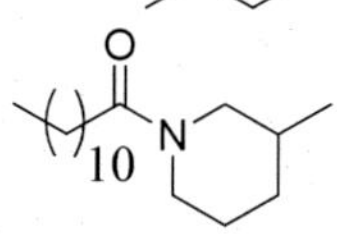
4a'	1-(1-cyclohexen-1-ylcarbonyl)piperidine (AI3-38739)	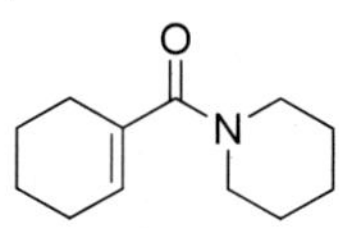
4b'	1-(cyclohexylcarbonyl) piperidine (AI3-36324)	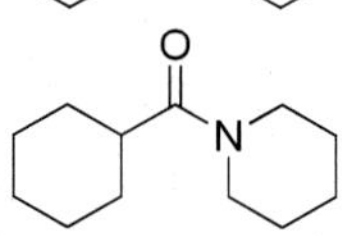
4c'	1-(cyclohexylcarbonyl)-3-methylpiperidine (AI3-36537)	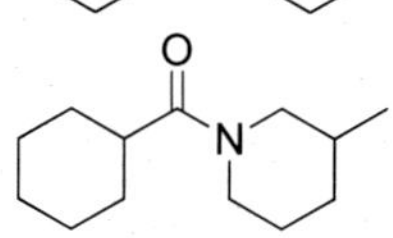
4d'	1-(cyclohexylcarbonyl)-4-methylpiperidine (AI3-36538)	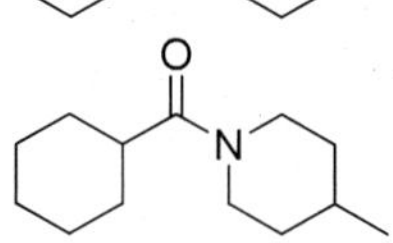
4e'	1-(3-cyclopentyl-1-oxopropyl)piperidine (AI3-38423)	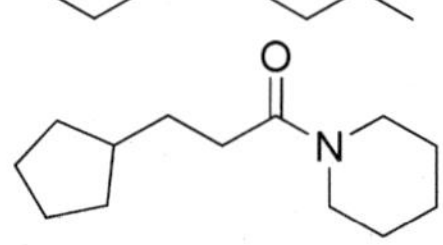

Continued on next page.

Table II. (Continued). Compounds used for the acylpiperidine repellent study

4f'[a]	1-(1-methylcyclohexylcarbonyl)-3-methylpiperidine	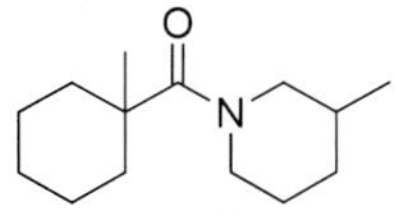
4g'	2-methyl-1-[(4-methylcyclohexyl) carbonyl]piperidine (AI3-39012)	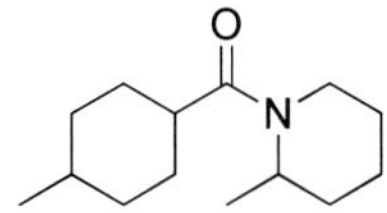
4h'	1-(cyclohexylcarbonyl)-2-ethylpiperididne (AI3-36539)	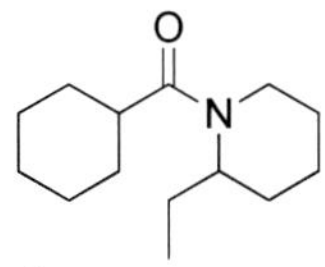
4i'	1-(cyclohexylacetyl)-2-methylpiperidine (AI3-37409)	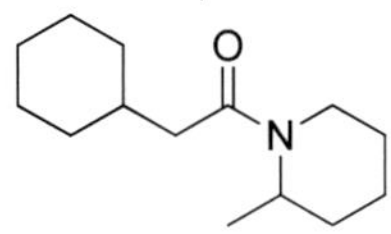
4j'[a]	1-(3-cyclohexyl-1-oxopropyl)-2-methylpiperidine (AI3-37424)	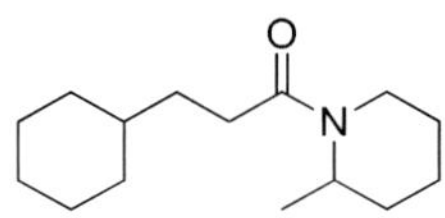
4k'	1-(3-cyclohexyl-1-oxopropyl)-3-methylpiperidine (AI3-37425)	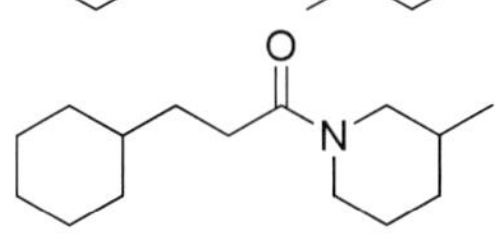
4l	1-(3-cyclohexyl-1-oxopropyl)-4-methylpiperidine	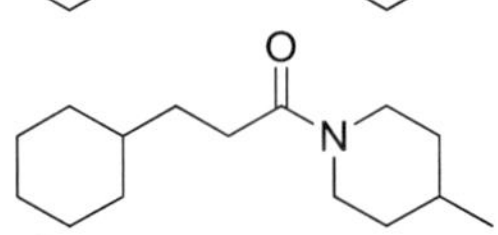
4m'[a]	1-(4-cyclohexyl-1-oxobutyl)-4-methylpiperidine	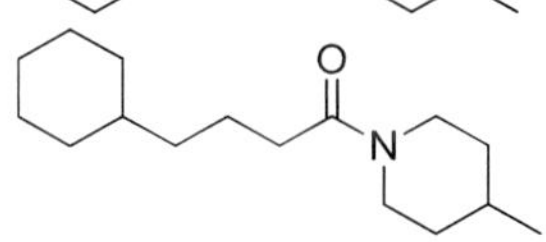
4n'[a]	1-(3-cyclopentyl-1-oxopropyl)-2-ethylpiperidine	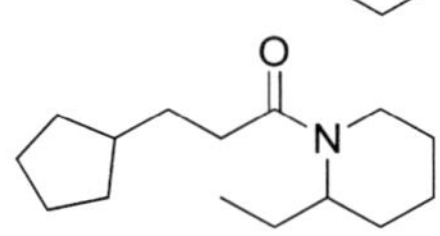
4o'[a]	1-(3-cyclohexyl-1-oxopropyl)-2-ethylpiperidine	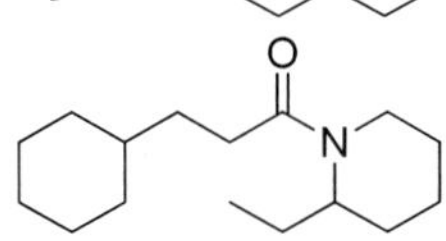
4p'[a]	1-(cyclohexylacetyl)-4-(phenylmethyl)piperidine	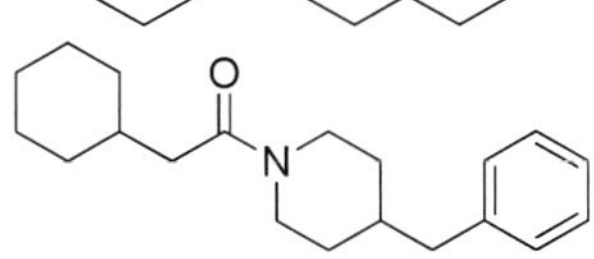
4q'[a]	1-(3-cyclohexyl-1-oxopropyl)-4-(phenylmethyl)piperidine	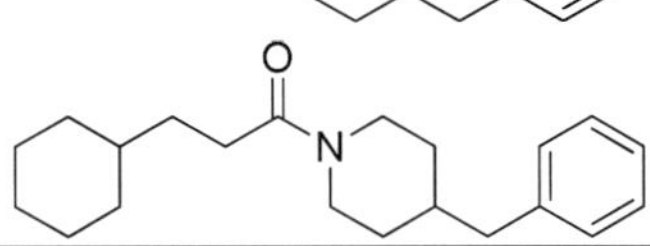

[a]Novel compounds

Bioassays of Compounds

The CPTs for the 34 acylpiperidines were determined at two selected concentrations (25 μmol/cm[2] and 2.5 μmol/cm[2]). At the higher concentration, approximately one-third of the compounds were repellent on cloth for a duration that was greater than three times the repellent duration of DEET (Figure 7). The compound 4-methyl-1-(1-oxo-10-undecylenyl)piperidine (**4n**) prevented bites for an average of 73 days compared to 17.5 days for DEET at the 25 μmol/cm[2] concentration (Figure 7, Table II). This same compound lasted an average of 13 days compared to 2.5 days for DEET when tested at the lower concentration.

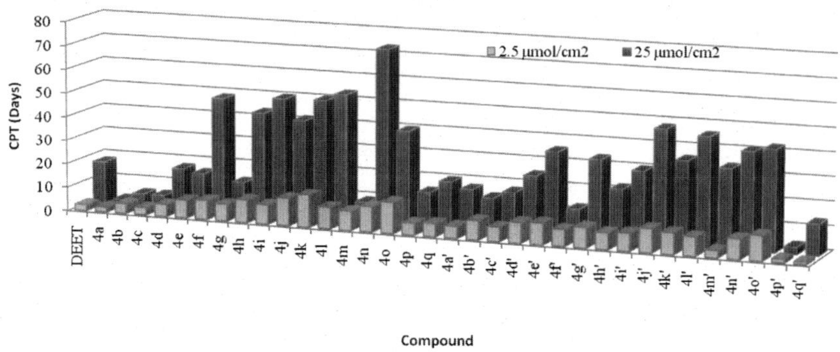

Figure 7. Complete protection time (CPT) of two concentrations of 23 novel and 11 previously tested acylpiperidines (see Table II for compound structures). (see color insert)

When the compounds that provided the greatest CPT are compared, there are noticeable similarities in their structures. Compound **4n** has a *para-* methyl on the piperidine ring with a 10-carbon terminally unsaturated chain as the acyl substituent. Very similar in structure to **4n** are **4k** and **4l**, which have the same acyl chain but no substituent on the piperidine ring (**4k**) and an *ortho-* ethyl on the piperidine ring (**4l**). Similarly, **4o** has a fully saturated 10-carbon acyl chain and again no substituent on the piperidine ring. Compounds **4i** and **4j** have 9-carbon fully saturated acyl chains with *ortho-* methyl on the piperidine for **4i** and a *para-* methyl on the piperidine for **4j**. Similar to **4j**, compound **4f** has a *para*-methyl on the piperidine ring, but instead has a 7-carbon saturated acyl chain. The cluster of compounds from **4j′-4o′** all have an acyl group consisting of a terminal cyclohexyl group or cyclopentyl in the case of **4n′**. The total number of carbons in the acyl group for each compound ranges from 7-9. The piperidine group either has a methyl substituent at the *ortho-*, *meta-*, or *para-* position, or has an *ortho-* ethyl group. Therefore, the general trend for acylpiperidines that last longer than DEET is that they: 1) contain no substituents, have monomethyl- or monoethyl- groups on the piperidine ring and 2) have an acyl group chain of 7-10 carbons, either saturated or terminally unsaturated, or having a terminal cyclopentane or cyclohexane. Presumably, the substituents reduce the volatility of these molecules and do not interfere with the structural properties that result in repellency when mosquitoes come in contact with these compounds.

If the repellency data for the 25 μmol/cm² and 2.5 μmol/cm² concentrations are converted to classes and plotted against the predicted class based on the ANN model, there appears to be little agreement between predicted and experimental classes for most of the compounds (Figure 8). In fact, the correlation between these data as predicted by ANN and the experimentally determined CPT converted to class is actually extremely low (R^2=0.007 and 0.006 for the high and low concentrations, respectively). Therefore classes are clearly not the best activity data to input for model development. Conversion back to classes results in non-linearity of the repellency data and reduces the number of "divisions" by which the repellent activity can be separated for the studied compounds.

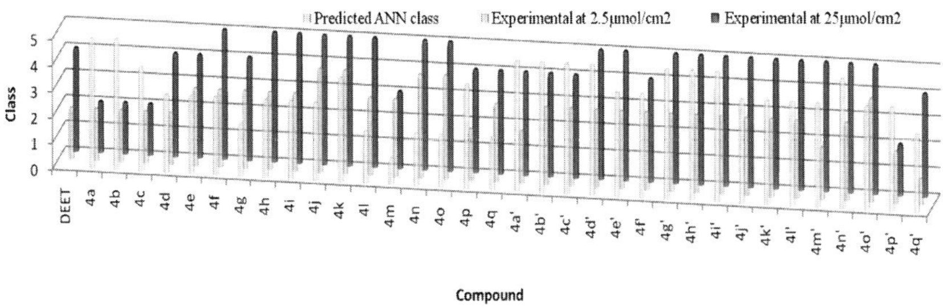

Figure 8. Comparison of predicted and experimental classes at two tested concentration levels of 25 and 2.5 μmol/cm² of 23 novel and 11 previously tested acylpiperidines.(see Table II for compound structures). (see color insert)

Maintaining activities as days of repellent duration is better for modeling purposes. Instead of an input if only 5 classes for the repellent activity, the range of activity can be input as the mean number of effective days of protection, from 1 to 73 days for the high concentration and from 0 to 13.5 days for the low concentration. Since each input was the mean duration of protection for two volunteers, this resulted in the possibility of half-day increments which effectively doubled the number of discrete values for the repellency activity at each concentration level.

Development of a QSAR Model

The results of bioassays (averaged days of CPT) were used to generate two QSAR models, one for the high (25 μmol/cm²) and one for the low (2.5 μmol/cm²) concentrations of compounds. Examination of the data distribution for each concentration revealed that the data acquired at the lower concentration had the more Gaussian distribution. In general, the more Gaussian the distribution of data used in a model, the more reliable the model is expected to be (*23*). The models were developed using 4 descriptors since adding additional descriptors complicated the model without adding a significant improvement in the predictive reliability (Table III).

Table III. Best 4 descriptors models and their statistical parameters. SOURCE: Reproduced from reference (*23*). Copyright 2008 The National Academy of Sciences of the USA

#	*B*[a]	*S*[b]	*t*[c]	*IC*[d]	*Name of descriptor*[e]
					25 μmol/cm²[f]
0	-188.8	84.08	-2.246		Intercept
1	-2686	461.3	-5.823	0.09647	Maximum 1-electron reactivity index for atom C
2	-2616	488.2	-5.359	0.7253	Principal moment of inertia C
3	2.040	0.6920	2.948	0.3632	Maximum e-e repulsion for bond C-C
4	-0.02195	0.009215	-2.382	0.7759	WPSA-2 Weighted PPSA (PPSA2*TMSA/1000)
					2.5 μmol/cm²[g]
0	-726.1	329.3	-2.205		Intercept
1	-68.13	9.393	-7.254	0.5248	YZ Shadow / YZ Rectangle
2	58.50	13.22	4.426	0.7120	Molecular volume/XYZ Box
3	-71.37	16.41	-4.350	0.5696	RNCG Relative negative charge (QMNEG/QTMINUS)
4	1.870	0.8053	2.321	0.2822	Minimum e-n attraction for bond C-O

[a] B, regression coefficient. [b] S, regression coefficient error. [c] *t*, Student criterion. [d] IC, partial intercorrelation. [e] PPSA, partial positively charged molecular surface area; WPSA, weighted PPSA; RNCG relative negative charge, ratio between the maximum atomic negative charge and sum of the negative atomic charges in the molecule. [f] N = 4; n = 34; $R^2 = 0.729$; $R^2_{cvOO} = 0.638$; $R^2_{cvMO} = 0.628$; F = 19.50; s = 9.769. [g] N = 4; n = 34; $R^2 = 0.689$; $R^2_{cvOO} = 0.608$; $R^2_{cvMO} = 0.582$; F = 16.05; s = 1.815.

Models for the high and low concentrations had good R^2 values (0.729 and 0.689, respectively) (Figure 9); however, it is obvious from Table III that the descriptors used to develop the models at each concentration were different. There are probably many reasons to explain these differences, but one of these is the difference in data distribution (normal or Gaussian), as noted earlier (*23*). Another reason may lie in the number of descriptors that the Codessa Pro software can employ in generating a model. Some of these descriptors may be similar to others and once the first is selected, other descriptors are chosen sequentially to be orthogonal to those already selected. An example of this similarity between non-identical descriptors can be seen in Table S5 from Katritzky et al. (*23*), where descriptor 3 (RNCG Relative Negative Charge) of the 2.5 μmol/cm² concentration model is highly intercorrelated with descriptors 2 (Principle Moment of Inertia C) and 4 (WPSA-2 weighted PPSA) of the 25 μmol/cm² concentration model at the 0.78 and 0.92 levels, respectively.

When the experimentally determined mean CPTs for the 25 μmol/cm² and the 2.5 μmol/cm² concentrations are compared to the QSAR model predicted values, it is visually evident that there is close agreement for many of the compounds (Figure 10 for the 25 μmol/cm2 and the Figure 11 for the 2.5 μmol/cm2) concentrations.

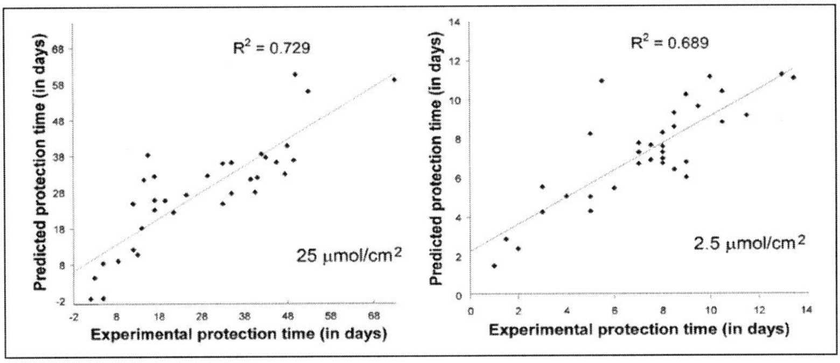

Figure 9. Comparison of predicted and experimental protection times for the two tested concentrations of acylpiperidines. Reproduced from reference (23). Copyright 2008 The National Academy of Sciences of the USA.

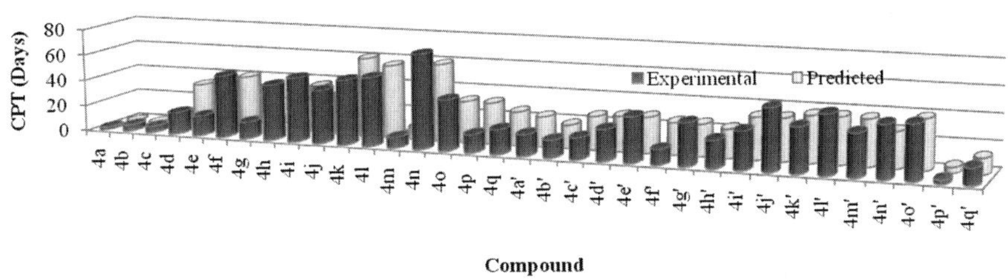

Figure 10. Comparison of experiment and predicted complete protection times(CPTs) for the high concentration (25 μmol/cm2) of 23 novel and 11 previously tested acylpiperidines (see Table II for compound structures). (see color insert)

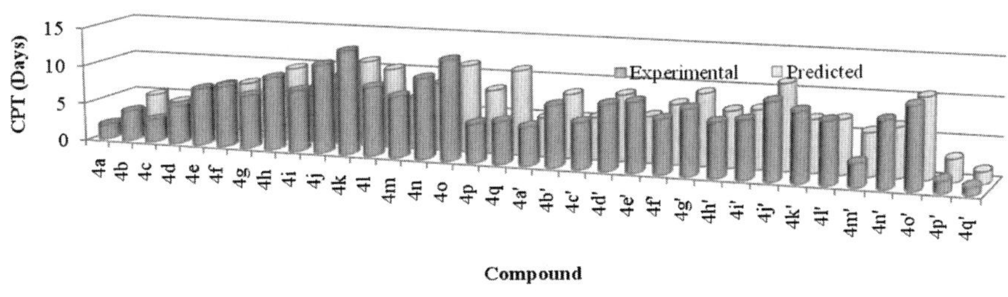

Figure 11. Comparison of experiment and predicted complete protection times(CPTs) for the low concentration (2.5 μmol/cm2) of 23 novel and 11 previously tested acylpiperidines (see Table II for compound structures). (see color insert)

Carboxamide Repellents

Encouraged by the success of modeling acylpiperidines using the days of CPT as the measured biological parameter, a more structurally diverse set of carboxamides was selected for ANN modeling to predict novel carboxamide structures as candidate repellents (*18*). The data were selected from repellency classes of compounds submitted to the USDA and archived between November, 1952, and November, 1992.

Artificial Neural Network Modeling

As in the acylpiperidine model development, classes of repellency were used for the carboxamides ANN model. A total of 167 carboxamides were randomly divided into a 120-compound training set and a 47-compound validation set. Up to 1557 descriptors were calculated for each of the compounds. The carboxamide ANN model differed from the acylpiperidines in both architecture and descriptors used. The architecture of the carboxamides model consisted of 6 input neurons, followed by 4 hidden neurons, with the final output neuron as the repellency class. The descriptors used for the input neurons were: 1) weighted partial positive surface area based on Zefirov's partial charge, 2) average H atom valency, 3) molecular volume/XYZ box, 4) highest normal mode vibration frequency, 5) highest normal mode vibration transition dipole, and 6) minimal resonance energy for the C-H bond.

The model predicted the correct class for 70 of the 120 compounds in the training set, with 115 out of 120 predicted within one class ($R^2 = 0.622$). The class 4 and class 5 compounds were used to design 144 similar structures that were input into the carboxamide ANN model. Of the 144 of these that were input, 34 of the compounds predicted to be the highest classes were then synthesized. Based on the structure of these compounds, 4 additional compounds were synthesized for bioassay testing (Table IV).

Synthesis

The selected 38 carboxamides **5a-l′** were synthesized according to the scheme in Figure 12 (*18*). Treatment of the carboxylic acids **1** with thionyl chloride and benzotriazole in methylene chloride in a 1:1:3 mole ratio at 20 °C gave 1-acylbenzotriazoles **2** using a modified procedure (*29*). Reaction of 1-acylbenzotriazoles **2** with one equivalent of secondary amines **4** either in THF at 20 °C or in toluene under reflux gave carboxamides **5a-5u, 5j′** and **5k′** in 70–100% and **5i′** and **5l′** in 36 and 28% yields respectively (path A) (*30*). Path B was chosen for the preparation of the carboxamides **5v-h′** to avoid undesired Michael-type addition of benzotriazole to carboxamides **5** when non-blocked α,β-unsaturated 1-acylbenzotriazoles **2** are reacted with a secondary amines under neutral conditions. The resulting mixture of by-product Bt1-adduct **6b**, byproduct Bt2-adduct **6a** and the desired product **5x** could not be separated by column chromatography. Acid chlorides **3** were either commercially available or prepared in situ by treatment of the corresponding carboxylic acids **1** with 20–27% excess

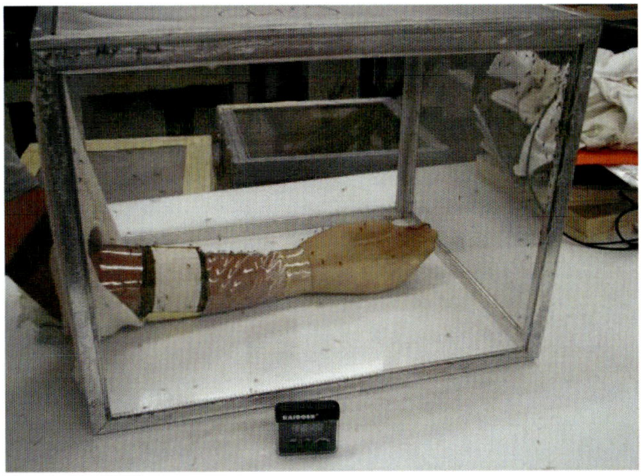

Figure 4. Photo Credit: Greg Allen, USDA-ARS. Reproduced from reference 18. Copyright 2010 Entomological Society of America. (Chapter 2)

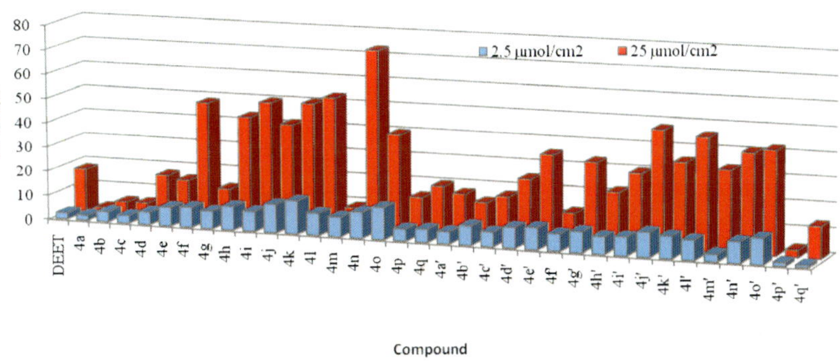

Figure 7. Complete protection time (CPT) of two concentrations of 23 novel and 11 previously tested acylpiperidines (see Table 2 for compound structures). (Chapter 2)

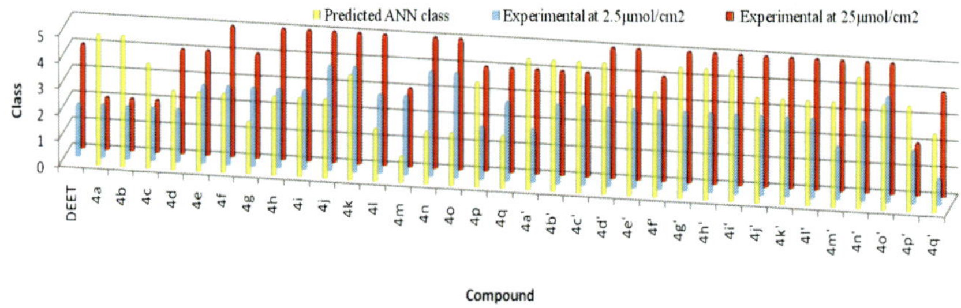

Figure 8. Comparison of predicted and experimental classes at two tested concentration levels of 25 and 2.5 µmol/cm² of 23 novel and 11 previously tested acylpiperidines. (see Table 2 for compound structures). (Chapter 2)

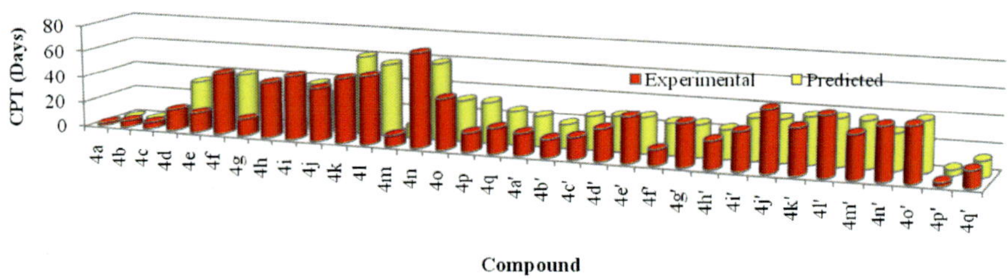

Figure 10. Comparison of experiment and predicted complete protection times(CPTs) for the high concentration (25 µmol/cm²) of 23 novel and 11 previously tested acylpiperidines (see Table 2 for compound structures). (Chapter 2)

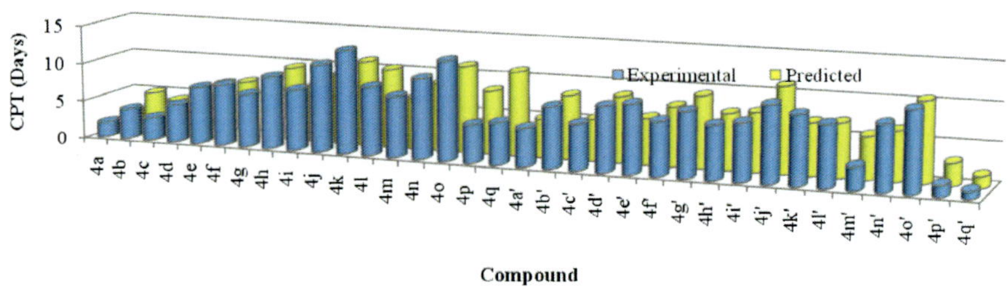

Figure 11. Comparison of experiment and predicted complete protection times(CPTs) for the low concentration (2.5 µmol/cm²) of 23 novel and 11 previously tested acylpiperidines (see Table 2 for compound structures). (Chapter 2)

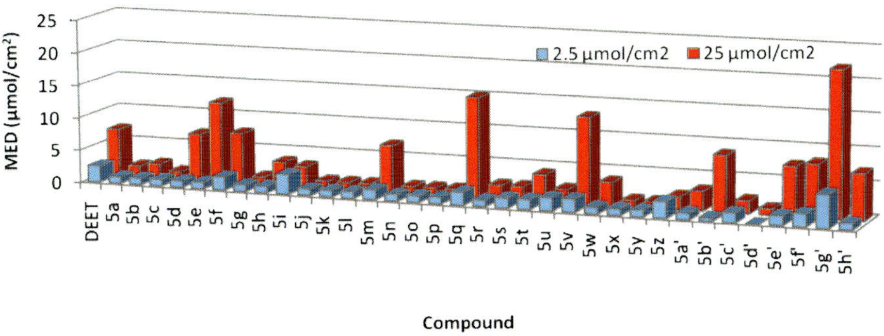

Figure 13. Complete protection time (CPT) of synthesized carboxamides compared to DEET at two concentrations (25 µmol/cm² and 2.5µmol/cm²). (Chapter 2)

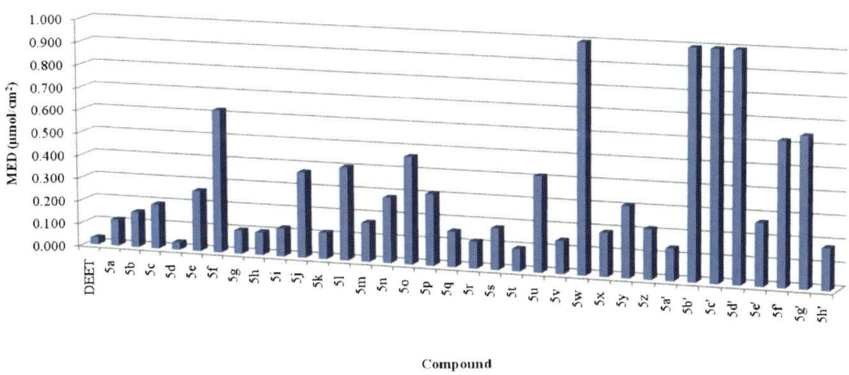

Figure 14. Minimum effective dosage (MED) of synthesized carboxamides compared to DEET. (Chapter 2)

of thionyl chloride at 20 °C overnight. Reaction of acid chlorides **3** with one equivalent of secondary amines in THF in the presence of 8% excess of sodium hydride at 0 to 20 °C led to formation of carboxamides **5v-h′** in 70–97% yield. The structures of the carboxamides **5a-l′** are given in Table IV.

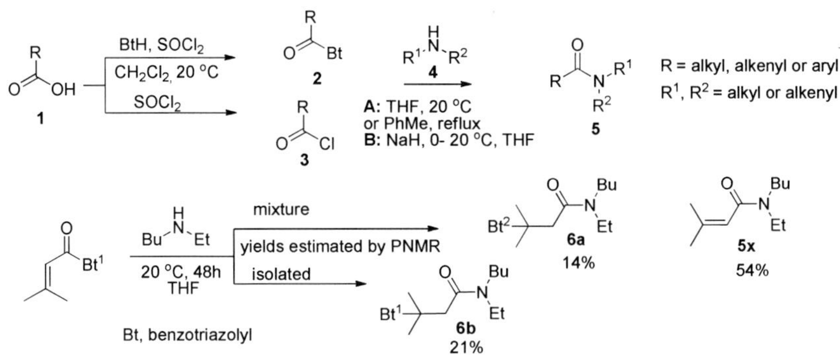

*Figure 12. Preparation of carboxamides **5**. Reproduced from reference (18). Copyright 2010 Entomological Society of America.*

Bioassays of Compounds

Although the ANN model adequately predicted classes for the compound structures used in the training and validation sets, when bioassayed most of the selected compounds were not as repellent as had been predicted. Possible reasons for this are that the diversity of the set and the non-linearity of the data prevented a successful correlation of predicted compounds with their experimentally determined efficacy. Over 50% of the compounds (23 out of 38) were predicted to be Class 4 and 5 (at least equivalent to DEET); however, only 11 of these had a CPT greater than that of DEET (Figure 13). At the 25 μmol/cm² concentration, the compound with the highest CPT (22 days), just over three times the duration of DEET, was a novel compound, (E)-*N*-cyclohexyl-*N*-ethyl-2-hexenamide (**5g′**) (Table IV). This compound lasted about twice as long as DEET when tested at the 2.5 μmol/cm² concentration.

Table IV. Compounds used for the carboxamide repellent study

ID	*Name*	*Structure*
DEET	N,N-diethyl-3-methylbenzamide	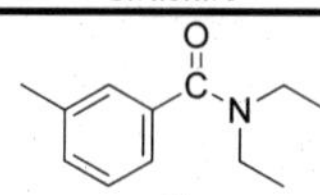
5[a]	N-butyl-N-methyl-hexanamide	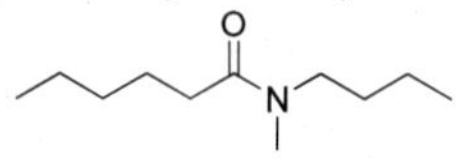
5b[a]	N-butyl-N-ethylhexanamide	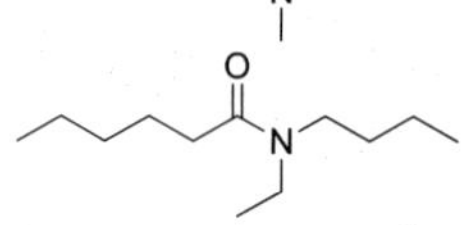
5c	N,N-diallylhexanamide	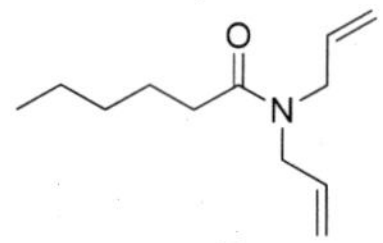
5d	hexahydro-1-(1-oxohexyl)-1H-azepine	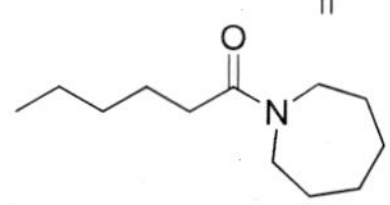
5e	N-cyclohexyl-N-ethylhexanamide	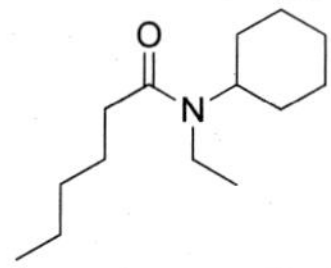
5f	N-ethyl-N-phenylhexanamide	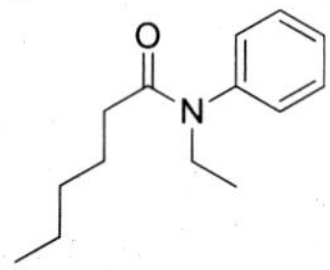
5g[a]	N-butyl-N-ethyl-2-methylpentanamide	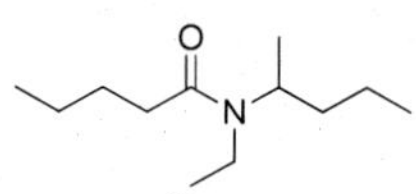
5h[a]	1-(1-azepanyl)-2-methyl-1-pentanone	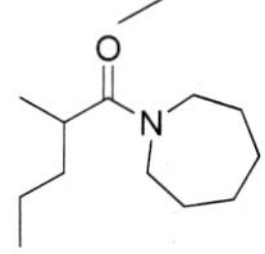
5i[a]	N-butyl-N,2-diethylbutanamide	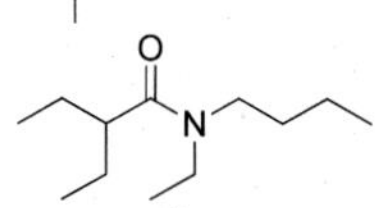
5j[a]	N,2-diethyl-N-(2-methyl-2-propenyl)butanamide	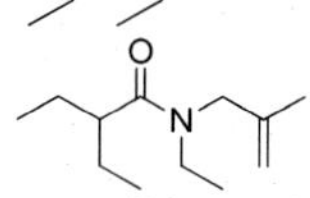
5k[a]	N-butyl-N-ethyl-3-methylbutanamide	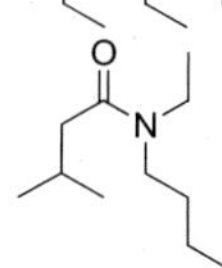
5l	N,N-diisobutyl-3-methylbutanamide	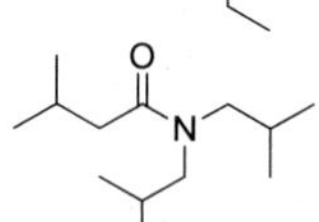

Continued on next page.

Table IV. (Continued). Compounds used for the carboxamide repellent study

Compound	Name	Structure
5m[a]	N-cyclohexyl-N-ethyl-3-methylbutanamide	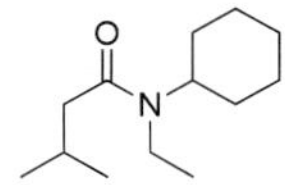
5n[a]	N-butyl-N-ethyl-2,2-dimethylpropanamide	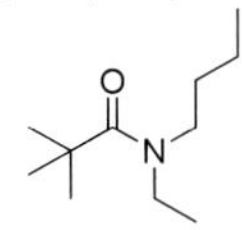
5o[a]	N-ethyl-2,2-dimethyl-N-(2-methyl-2-propenyl)propanamide	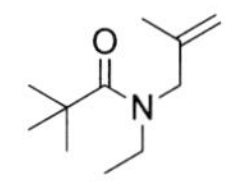
5p	1-(1-azepanyl)-2,2-dimethyl-1-propanone	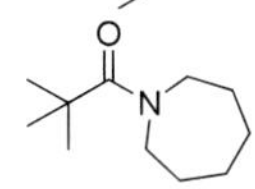
5q[a]	N-butyl-N-ethyl-2-methylbenzamide	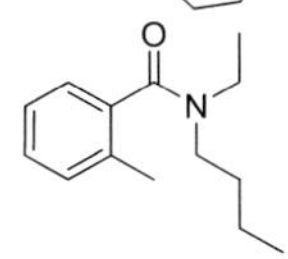
5r[a]	(E)-N-butyl-N-ethyl-2-methyl-2-pentenamide	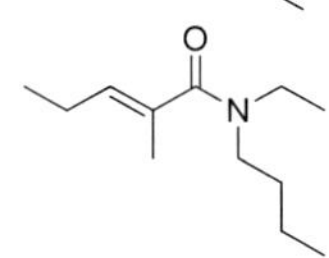
5s[a]	(E)-N-ethyl-2-methyl-N-(2-methyl-2-propenyl)-2-pentenamide	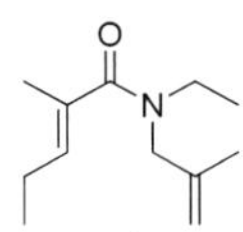
5t[a]	(E)-1-(1-azepanyl)-2-methyl-2-penten-1-one	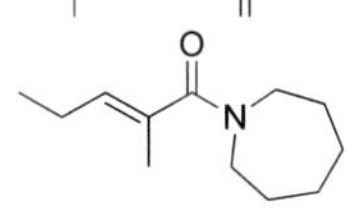
5u	(E)-2-methyl-N,N-di-2-propenyl-2-pentenamide	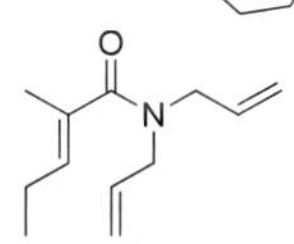
5v[a]	N-ethyl-2-methyl-N-(2-methyl-2-propenyl)benzamide	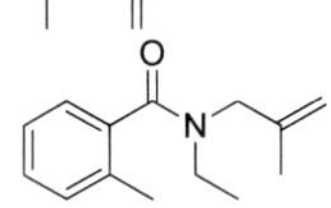
5w	N-ethyl-2-methyl-N-phenyl-benzamide	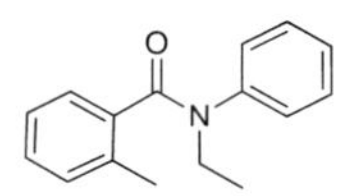
5x[a]	N-butyl-N-ethyl-3-methyl-2-butenamide	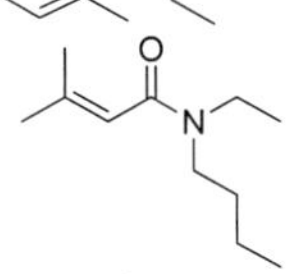
5y[a]	N-ethyl-3-methyl-N-(2-methyl-2-propenyl)-2-butenamide	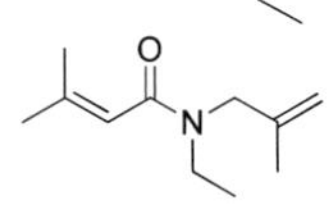

Continued on next page.

Table IV. (Continued). Compounds used for the carboxamide repellent study

5z	N,N-diisobutyl-3-methyl-crotonamide	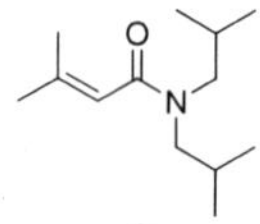
5a'	hexahydro-1-(3-methylcrotonoyl)-1H-azepine	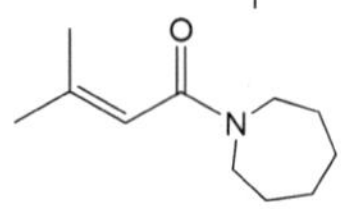
5b'	N-butyl-N-ethyl-cinnamamide	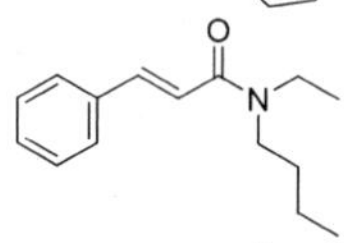
5c'[a]	N,N-bis(2-methylpropyl)-3-phenyl-2-propenamide	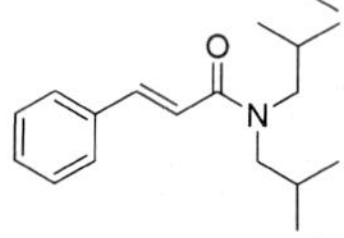
5d'	N-ethyl-N,3-diphenyl-2-propenamide	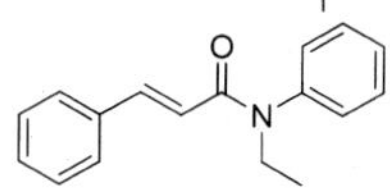
5e'[a]	(E)-N-n-butyl-N-ethyl-2-hexenamide	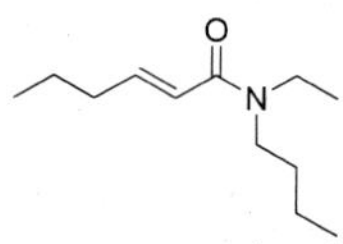
5f'[a]	(E)-N,N-di-(2-methylpropyl)-2-hexenamide	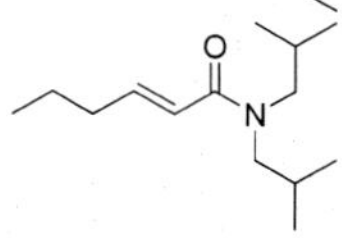
5g'[a]	(E)-N-cyclohexyl-N-ethyl-2-hexenamide	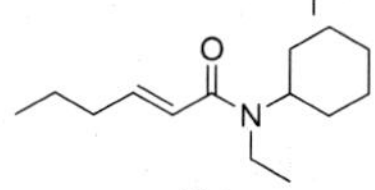
5h'[a]	N-butyl-N-methyl-5-hexynamide	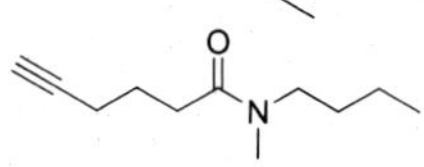
5i'[a]	N,3-dicyclohexyl-N-ethylpropanamide	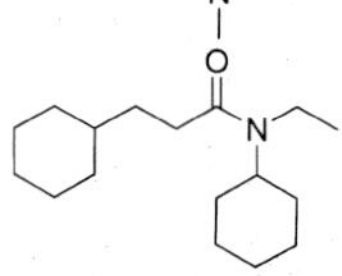
5j'[a]	(E)-N,2-dimethyl-N-octylpent-2-enamide	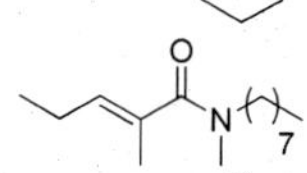
5k'[a]	N-cyclohexyl-N-methylheptanamide	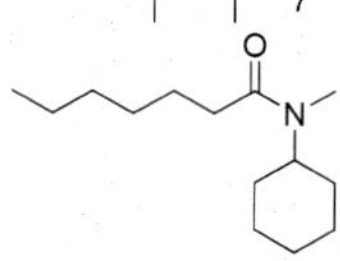
5l'[a]	(E)-N-cyclohexyl-N-ethyl-2-methylpent-2-enamide	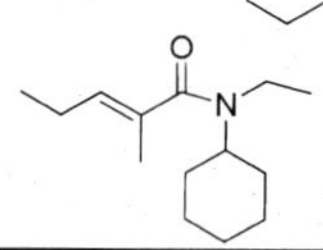

[a]Novel compounds

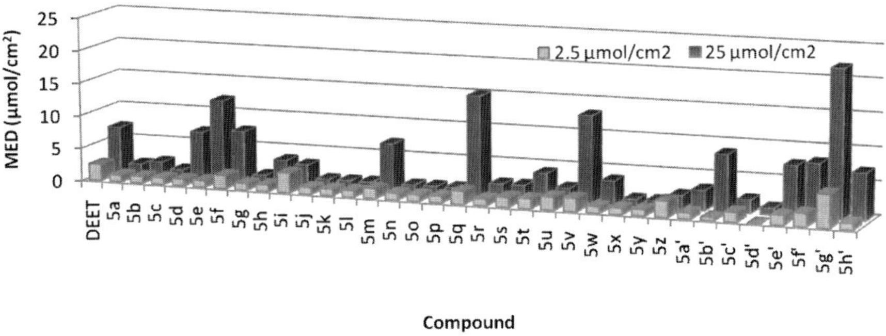

Figure 13. Complete protection time (CPT) of synthesized carboxamides compared to DEET at two concentrations (25 µmol/cm^2 and 2.5µmol/cm^2). (see color insert)

Unlike the acylpiperidines, similarities among the best repellents (those with highest CPT) are not as apparent. Compound **5g′** has an ethyl- group and cyclohexyl group attached to the amide nitrogen, with a 5-carbon chain in the acyl group that is unsaturated next to the carbonyl. The same substituents on the *N*- and a fully saturated 5-carbon acyl chain results in **5e** having a CPT about two times longer than DEET. Compound **5q** had the second longest CPT and its structure is similar to DEET, having an *ortho* - methylbenzene attached to the carbonyl carbon, and ethyl and butyl groups attached to the nitrogen. Another of the better compounds, **5v**, is similar to **5q** on the acyl side, also contains an ethyl group attached to the amide nitrogen, but has a 2-methylpropene as the other substituent.

The non-linearity of the data and lack of widespread differences in repellent duration did not allow the development of QSAR models (*18*). Therefore, it was decided to examine the MED of the synthesized carboxamides. The compounds hexahydro-1-(1-oxohexyl)-1H-azepine (**5d**) had a MED that was equivalent to that of DEET (0.047 ± 0.007) µmol/cm^2 (Figure 14). Other compounds that were nearly equivalent in potency were (E)-1-(1-azepanyl)-2-methyl-2-penten-1-one (**5t**) at 0.098 ± 0.20 µmol/cm^2 and similarly structured 1-(1-azepanyl-)-2-methyl-1-pentanone (**5h**) at 0.102 ± 0.033 µmol/cm^2, followed by *N*-butyl-*N*-ethyl-2-methylpentanamide (**5g**) at 0.104 ± 0.16 µmol/cm^2. There was no apparent correlation noticeable between the most potent compounds having the lowest MED and compounds that were the least volatile (greatest CPT). However, it appears that the most potent repellents (those having the lowest MED) contain an azepine ring on the amide nitrogen. The compounds **5d**, **5h**, **5t** all have 5-carbon chains on the acyl side, with **5h** and **5t** having a methyl branch and **5t** with an unsaturated bond. The compound **5a′** has a relatively low MED and is similar to **5t** except that the unsaturated acyl group contains one less carbon. The least potent of this series is **5p**, containing a *t*-butyl group.

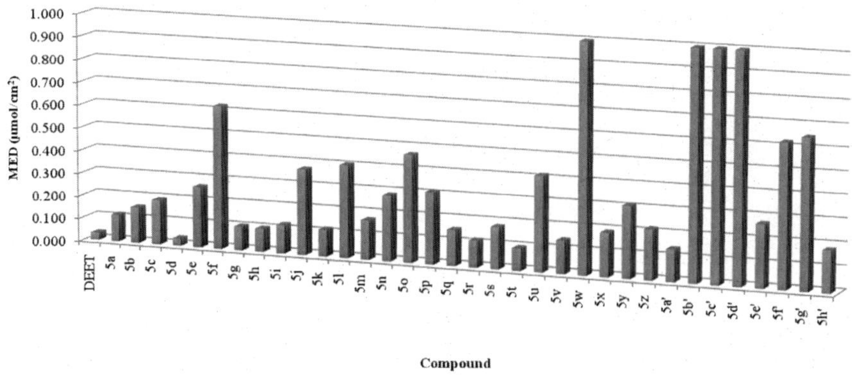

Figure 14. Minimum effective dosage (MED) of synthesized carboxamides compared to DEET. (see color insert)

Summary and Future Work

The repellency class data of a set of acylpiperidines from the USDA archive were used to develop suitable ANN models to predict new repellent structures. Predicted compounds that had not been previously examined for repellency along with compounds tested as repellents during the past 70 years were bioassayed for CPT. The results were used to develop a successful QSAR model to predict repellency duration (CPT) giving excellent correlation with experimental data. Some of these compounds had a duration of repellency three times better than DEET.

The approach used to produce the successful modeling and prediction of acylpiperidines was also applied to a subset of carboxamides. Perhaps due to the greater structural diversity, or imprecision in the non-linear class data, ANN models were not as successful in the prediction of repellents with high efficacy. However, despite the inability to produce a QSAR model of the carboxamide, about one-third of them had a CPT comparable or superior to DEET and another of the compounds had a MED equivalent to DEET.

Ongoing studies are being conducted to evaluate the acylpiperidines and carboxamides against other species, specifically ticks, and mosquito species that transmit malaria, such as *Anopheles gambiae and An. albimanus*. Traditionally, these mosquito species have been more difficult to repel than *Ae. aegypti*. Additionally, modeling approaches are being applied to mosquito and house fly adulticide and larvicide data found in the USDA archive.

Acknowledgments

We appreciate contributions to this work by A. Katritzky, S. Slavov, D. Dobchev, Z. Wang, C.D. Hall, N. Newlon, G. Allen, N. Elejalde, G. Clark, and K. Linthicum, and to those who made innumerable contributions to the USDA screening program since 1942. This study was partly supported by the Deployed War-Fighter Protection Research Program and funded by the U.S. Department of Defense through the Armed Forces Pest Management Board.

References

1. Patterson, G. M. *The Mosquito Wars: A History of Mosquito Control in Florida*; University Press of Florida: Gainesville, FL, 2004; pp 95–100.
2. *Results of Screening Tests with Materials Evaluated as Insecticides, Miticides and Repellents at the Orlando, Florida, Laboratory, April, 1942 to April, 1947*; Publication E-733; Bureau of Entomology and Plant Quarantine, United States Department of Agriculture (USDA): Washington, DC, 1947; pp 1–13.
3. Knipling, E. F. *Insect Control Investigations of the Orlando, Florida, Laboratory During World War II*; Smithsonian Report for 1948 (Publication 3968); U.S. Government Printing Office: Washington, DC, 1949; pp 331–348.
4. *Quarterly Report, Entomological Research, Quarter Ending June 30, 1952*; Bureau of Entomology and Plant Quarantine, United States Department of Agriculture (USDA): Washington, DC, 1952; pp 1–16.
5. *Quarterly Report, Entomological Research, Quarter Ending March 31, 1953*; Bureau of Entomology and Plant Quarantine, United States Department of Agriculture (USDA): Washington, DC, 1953; p 13.
6. Höltje, H.-D.; Sippl, W.; Rognan, D.; Folkers, G. *Molecular Modeling, Basical Principles and Applications*, 3rd ed.; Wiley-VCH Verlag GmbH & Co., KGaA: Weinheim, Germany, 2008; pp 2–8.
7. Gertler, S. I. U.S. Patent 2,408,389, 1946.
8. McCabe, E. T.; Barthel, F. W.; Gertler, S. I.; Hall, S. A. Insect repellents. III. *N,N*-diethylamides. *J. Org. Chem.* **1954**, *19*, 493–498.
9. Frenkel, D.; Smit, B. *Understanding Molecular Simulation: From Algorithms to Applications*; Academic Press: San Diego, CA, 1996, pp 1–6.
10. Hansch, C.; Fujita, T. In *Classical and Three-Dimensional QSAR in Agrochemistry*; Hansch, C., Fujita, T., Eds.; ACS Symposium Series 606; American Chemical Society: Washington, DC, 1995, pp 1–12.
11. Moore, S. J.; Debboun, M. In *Insect Repellents Principles, Methods, and Uses*; Debboun, M., Frances, S. P, Strickman, D., Eds.; CRC Press: Boca Raton, FL, 2007, pp 3–29.
12. Ma, D.; Bhattacharjee, A. K.; Gupta, R. K.; Karle, J. M. *Am. J. Trop. Med. Hyg.* **1999**, *60*, 1–6.
13. Gupta, R. K.; Bhattacharjee, A. K. In *Insect Repellents Principles, Methods, and Uses*; Debboun, M., Frances, S. P, Strickman, D., Eds.; CRC Press: Boca Raton, FL, 2007, pp 195–228.

14. Bhonsle, J. B.; Bhattacharjee, A. K.; Gupta, R. K. *J. Mol. Model.* **2007**, *13*, 179–208.
15. Suryanarayana, M. V. S.; Pandey, K. S.; Prakash, S.; Raghuveeran, C. D.; Dangi, R. S.; Swamy, R. V.; Rao, K. M. *J. Pharm. Sci.* **1991**, *80*, 1055–1057.
16. Katritzky, A. K.; Dobchev, D. A.; Tulp, I.; Karelson, M.; Carlson, D. A. *Bioorg. Med. Chem. Lett.* **2006**, *16*, 2306–2311.
17. Codessa Pro Software, University of Florida, 2002. www.codessa-pro.com.
18. Katritzky, A. R.; Wang, Z.; Slavov, S.; Dobchev, D. A.; Hall, C. D.; Tsikolia, M.; Bernier, U. R.; Elejalde, N. M.; Clark, G. G.; Linthicum, K. J. *J. Med. Entomol.* **2010**, *47*, 924–938.
19. Hanrahan, G. *Artificial Neural Networks in Biological and Environmental Analysis*, 1st ed.; CRC Press: Boca Raton, FL, 2011; p 3.
20. Posey, K. H.; Schreck, C. E. *Mosq. News* **1981**, *41*, 566–568.
21. *Repellent Activity of Compounds Submitted by the Walter Reed Army Institute of Research. Part I. Protection Time and Minimum Effective Dosage Against Aedes Aegypti mosquitoes*; U.S. Department of Agriculture Technical Bull. 1549; United States Department of Agriculture (USDA): Washington, DC, 1977; pp 1−2.
22. Barnard, D. R.; Bernier, U. R.; Xue, R.-D.; Debboun, M. In *Insect Repellents Principles, Methods, and Uses*; Debboun, M., Frances, S. P, Strickman, D., Eds.; CRC Press: Boca Raton, FL, 2007, pp 103−109.
23. Katritzky, A. R.; Wang, Z.; Slavov, S.; Tsikolia, M.; Dobchev, D.; Akhmedov, N. G.; Hall, C. D.; Bernier, U. R.; Clark, G. G.; Linthicum, K. J. *Proc. Nat. Acad Sci. (U.S.A.)* **2008**, *105*, 7359–7364.
24. McGovern, T. P.; Schreck, C. E.; Jackson, J. *Mosq. News* **1978**, *38*, 346–349.
25. Basak, S. C.; Natarajan, R.; Nowak, W.; Miszta, P.; Klun, J. A. *SAR QSAR Environ. Res.* **2007**, *18*, 237–250.
26. Klun, J. A.; Shmidt, W. F.; Debboun, M. *J. Med. Entomol.* **2001**, *38*, 809–812.
27. McGovern, T. P; Schreck, C. E. U.S. Patent 4,291,041, 1981.
28. Katritzky, A. R.; Suzuki, K.; Wang, Z. *Synlett* **2005**, *11*, 1656–1665.
29. Katritzky, A. R.; Cai, C.; Singh, S. K. *J. Org. Chem.* **2006**, *71*, 3375–3380.
30. Katritzky, A. R.; He, H. Y.; Suzuki, K. *J. Org. Chem.* **2000**, *65*, 8210–8213.

Chapter 3

Callicarpenal and Intermedeol: Two Natural Arthropod Feeding Deterrent and Repellent Compounds Identified from the Southern Folk Remedy Plant, *Callicarpa americana*

Charles L. Cantrell[*,1] **and Jerome A. Klun**[2]

[1]**USDA-ARS, Natural Products Utilization Research Unit, University, Mississippi 38677, U.S.A.**
[2]**(Retired), USDA-ARS, Beltsville Agricultural Research Center, Invasive Insects Biocontrol and Behavior Laboratory, Beltsville, Maryland 20705, U.S.A.**
[*]**E-mail: charles.cantrell@ars.usda.gov**

In previous studies on the American beautyberry (*Callicarpa americana*), it was demonstrated that callicarpenal and intermedeol were responsible for the arthropod repellent and feeding deterrent activity of this folk remedy. Both compounds showed significant bite-deterring activity against *Aedes aegypti* and *Anopheles stephensi*. Callicarpenal and intermedeol were evaluated in laboratory bioassays for repellent activity against host-seeking nymphs of the blacklegged tick, *Ixodes scapularis*, a vector for Lyme disease, against *Amblyomma americanum*, a vector for erlichiosis, and *Amblyomma cajennense*, a vector for Rocky Mountain spotted fever. Callicarpenal and intermedeol were also evaluated for repellency using multiple choice digging bioassays against workers of red imported fire ants, *Solenopsis invicta*, black imported fire ants, *Solenopsis richteri,* and a hybrid of the two species. Chemical modifications were performed on callicarpenal in a preliminary structure–activity relationship study against *Ae. aegypti*. In continuation with this study, callicarpenal diethyl amine and piperidine analogs were synthesized and evaluated against both *Ae. aegypti* and *Ae. albopictus* at 25 nmoles/cm^2. We also conducted studies to determine the optimal extraction conditions for obtaining callicarpenal from dry leaves, and these methods were used to evaluate the variation in the dry weight concentration of callicarpenal at various stages of plant development throughout a single growing season.

Introduction

An analysis of worldwide pesticide sales indicates that herbicides account for 45.4% of the agrochemical market, followed by insecticides 27.5%, fungicides 21.7% and other products 5.4% (*1*). Other recent reports on worldwide sales of insecticides have continued to indicate an increase in natural product and natural product-derived insecticide sales while sales of organophosphates are predicted to decline with further restrictions. One of the reports (*2*) indicates that five groups of insecticides (carbamates, neonicotinoids, pyrethroids, organophosphates, and natural products) accounted for over three-quarters of worldwide sales. It is important to note that three of these groups are either completely natural product based or derived from natural products. Their combined worldwide sales accounted for 42.8% with the pyrethroids at 19.5%, neonicotinoids at 15.7%, and natural products at 7.6%.

Insect and tick repellents in the market continue to be dominated by the active ingredient, *N,N*-diethyl-*m*-toluamide (DEET), since its discovery in the 1950's. At present, approximately 140 products containing DEET are registered with the Environmental Protection Agengy (EPA) by about 39 different companies, and the U.S. EPA estimates that more than 38% of U.S. population uses a DEET-based repellent every year.

Additional synthetic based insect repellent active ingredients (AI) such as 2-(2-hydroxyethyl)-1-piperidinecarboxylic acid 1-methylpropyl ester (picaridin) and 3-[*N*-butyl-*N*-acetyl]-aminopropionic acid (IR3535) continue to increase their market share worldwide. Inspection of the structures of DEET, picaridin, and IR3535 (Figure 1) reveals a common diethyl amide structural motif or functional group present in all three of these widely available AI's. Surprisingly, there is limited chemical diversity available to consumers even when choosing effective synthetic insect repelling products.

Fortunately, the market is beginning to respond to consumer demand for natural product based insect repellents as alternatives to the more readily available synthetic AI based products. Such products commonly contain plant essential oils as active ingredients. One in particular, oil of lemon eucalyptus has shown good promise due in part to the activity of its *p*-menthane-3,8-diol (PMD) active ingredient. Unfortunately, reports of eye irritation from using PMD exist (*3*) but not in the peer-reviewed literature.

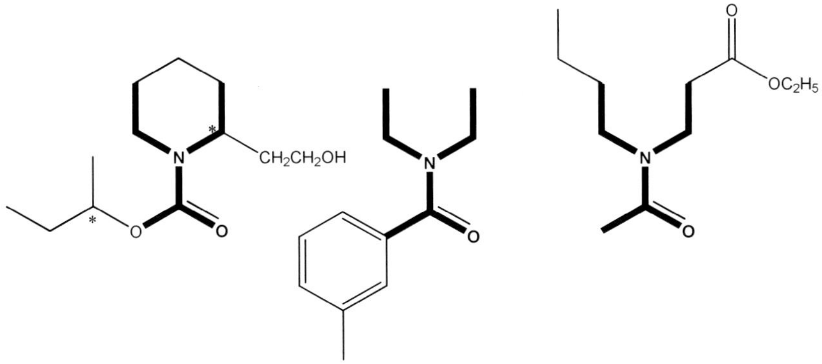

Picaridin DEET IR3535

Figure 1. Chemical structures of common commercially available active repellent ingredients with bold highlights indicating similar structural motifs and/or functional groups.

Callicarpenal and Intermedeol Repellency Studies

In Mississippi, crushed leaves of American beautyberry, *Callicarpa americana* L. (Verbenaceae), were placed under the harnesses of draft animals as a traditional means to protect the animals from hematophagous insects (*4*, *5*). Beautyberry leaves have been used as recently as the 1980s to repel arthropods (Charles Bryson, personal communication). Specific identification of the compounds responsible for the mosquito biting deterrency in the leaves of this folk remedy was recently completed (*4*). Briefly, a bioassay-directed fractionation approach was used in the study and the targeted arthropod was the mosquito *Aedes aegypti*, the vector for the yellow fever virus. The bioassay system utilized was the K & D Module (*4*).

The bioassay-directed fractionation approach began with a series of crude extracts of the plant leaves and indicated that the essential oil held the highest concentration of biting-deterrent constituents. Ultimately, the study identified the compounds callicarpenal and intermedeol (Figure 2) as those responsible for the biting-deterrency of the leaves and hence the folk remedy. A typical GC-MS chromatogram is shown in Figure 3. Bioactive compounds callicarpenal and intermedeol are labeled.

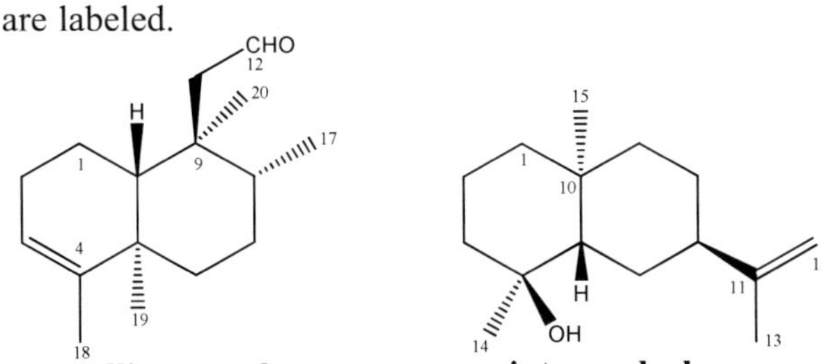

callicarpenal **intermedeol**

Figure 2. Chemical structures of compounds isolated from C. americana.

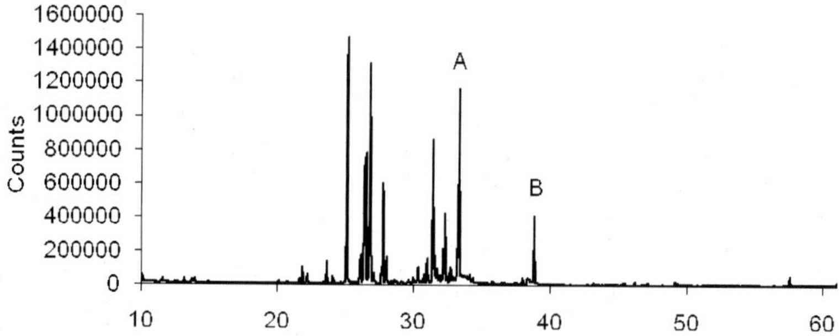

Figure 3. GC-MS total ion chromatogram for C. americana essential oil extracts (A = intermedeol; B = callicarpenal).

Following the identification of intermedeol and callicarpenal as the bioactive constituents in *C. americana*, both compounds were evaluated against both *Ae. aegypti* and *Anopheles stephensi* at 25 nmoles/cm^2 in the K & D bioassay (Figure 4). Briefly, the K & D bioassay consists of wells containing human blood cells in a water-bath warmed (38°C) reservoir and covered with a collagen membrane. The blood-membrane unit simulates a human host for mosquito feeding. The bioassay consisted of callicarpenal and intermedeol as the test subjects, (1*S*, 2′*S*)-2-methylpiperidinyl-3-cyclohexen-1-carboxamide (SS220) as a positive control and ethanol as the solvent control. The bioassay was performed as previously described by Cantrell et. al. (*4*). SS220 is a piperidine analog and commonly used with the K & D Module bioassay system as a positive control (*4*). Against *Ae. aegypti*, callicarpenal and intermedeol had significant activity and were only slightly less effective than SS220, which were both equally active against *Ae. aegypti*. Against *An. stephensi,* callicarpenal and intermedeol were as effective as SS220.

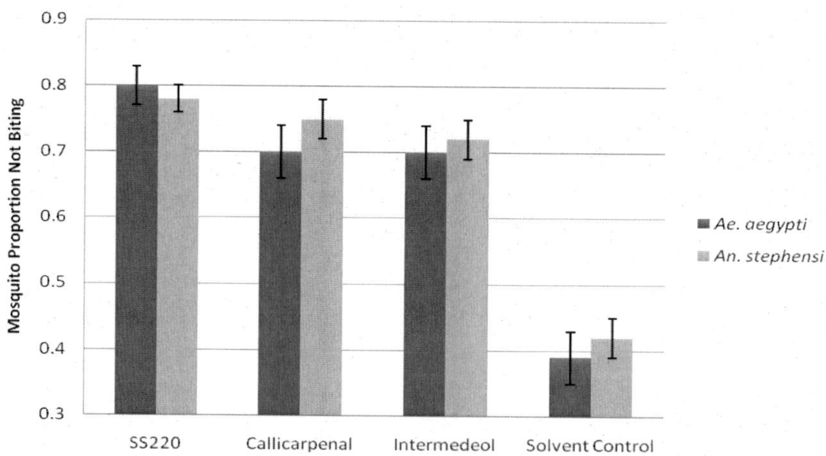

Figure 4. C. americana isolated compounds at 25 nmole/cm^2 vs. Ae. aegypti and An. stephensi.

As an extension to the above evaluation, callicarpenal was evaluated in the same K & D Module bioassay system with DEET as the positive control (*6*) against both *Ae. aegypti* and *An. stephensi* at 25 nmoles/cm^2 (Figure 5). The proportion of mosquitoes not biting shows that both compounds were significantly more effective than control ethanol-treated cloth in deterring biting. Callicarpenal deterred the biting of both mosquito species more effectively than DEET in this 3 minute bioassay. This evidences an outstanding performance of the plant derived compound callicarpenal over DEET.

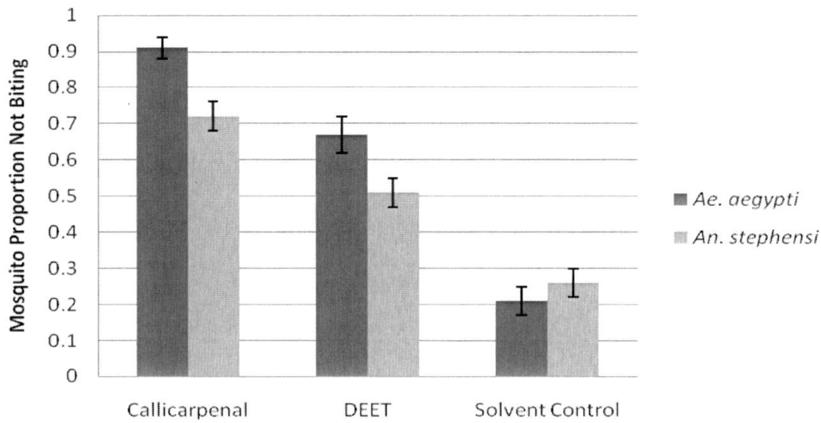

Figure 5. Callicarpenal and DEET at 25 nmoles/cm^2 versus Ae. aegypti and An. stephensi.

Since the original report on callicarpenal and intermedeol in 2005, many reports have appeared in the literature regarding the arthropod repelling effectiveness of these compounds in addition to a structure−activity relationship study. Carrol et. al. evaluated callicarpenal and intermedeol in laboratory bioassays for repellent activity against host-seeking nymphs of the blacklegged tick, *Ixodes scapularis*, and lone star tick, *Amblyomma americanum* (*7*, *8*). Callicarpenal and intermedeol, at 155 nmole/cm^2 cloth repelled 98 and 96% of *I. scapularis* nymphs, respectively. Dose response tests with *I. scapularis* nymphs showed no difference in repellency among callicarpenal, intermedeol and DEET. Callicarpenal, at 155 nmole/cm^2 cloth, repelled 100 and 53.3% of *I. scapularis* nymphs at 3 and 4 h, respectively, after the cloth was treated, whereas intermedeol repelled 72.5% of *I. scapularis* nymphs 3 h after treatment. Neither compound was very effective at repelling *A. americanum*.

Chen et. al. evaluated callicarpenal and intermedeol repellency using multiple choice digging bioassays against workers of red imported fire ants, *Solenopsis invicta,* black imported fire ants, *Solenopsis richteri,* and a hybrid of the two (*9*). Callicarpenal showed significant repellency at concentration as low as 50 ppm against both red imported fire ant colonies and 6.25 ppm against all black imported fire ant colonies and hybrid colonies. Intermedeol showed significant repellency at concentration as low as 1.50 ppm against both red imported fire ant colonies and 6.25 ppm against all black imported fire ant colonies and hybrid colonies.

Soares et. al. evaluated callicarpenal and intermedeol in laboratory bioassays for repellent activity against host-seeking nymphs of the tick *Amblyomma cajennense* (*10*) using a fingertip bioassay. Callicarpenal and intermedeol showed a lower effectiveness than DEET (*11*), as verified by the higher effective concentrations (ECs), shorter duration and lower percentage of repellency. Briefly, the EC_{50} for callicarpenal was 0.084 mg/cm^2, the most active of any of the botanicals evaluated in the manuscript. Intermedeol was also evaluated and gave an EC_{50} of 0.583 mg/cm^2.

Callicarpenal Structure−Activity Relationship Studies

Cantrell et. al. performed structural modifications on callicarpenal in an effort to understand the functional groups necessary for maintaining and/or increasing its biting deterrent activity against *Ae. aegypti* and to possibly lead to more effective insect control agents (*12*). All modifications in the study targeted the C-12 aldehyde or the C-3 alkene functionalities or combinations thereof. Mosquito biting deterrency appeared to be influenced most by C-3 alkene modification as evidenced by catalytic hydrogenation that resulted in a compound having significantly less effectiveness than callicarpenal at a test amount of 25 nmol/cm^2. Oxidation and/or reduction of the C-12 aldehyde did not diminish mosquito biting deterrency, but, at the same time, none of the modifications were more effective than the parent compound callicarpenal in deterring mosquito biting.

More recently, modifications to callicarpenal have continued to target the C-12 aldehyde with efforts aimed at the production of biting deterrent amides. The importance of the amide functional group to the activity of commercial insect repellents such as DEET and picaridin was mentioned in the introduction and highlighted in Figure 1. What we have tried to introduce into the callicarpenal structure via a condensation reaction of callicarpenoic acid are the diethyl amine and piperidine functional groups that exist in both DEET and picaridin, respectively (Figure 6). Synthetic analogs were produced from this two step process in yields of 22% and 27% for the diethyl amine analog and piperidine analogs, respectively.

Callicarpenal and its diethyl amine and piperidine analogs were evaluated against both *Ae. aegypti* and *Ae. albopictus* at 25 nmoles/cm^2 in the K & D bioassay model (Figure 7). The bioassay consisted of callicarpenal as a positive control, its diethyl amine and piperidine analogs as the test compounds, and acetonitrile as the solvent control. The bioassay was performed essentially as previously described by Cantrell et. al. (*4*) with 3 minutes of exposure. Against *Ae. aegypti*, both the diethyl amine and piperidine analogs of callicarpenal had significant activity however both compounds were less effective than callicarpenal and the piperidine analog was more effective than the diethyl amine analog. Against *Ae. albopictus,* again both the diethyl amine and piperidine analogs of callicarpenal had significant activity and both compounds were equivalent to each other and callicarpenal.

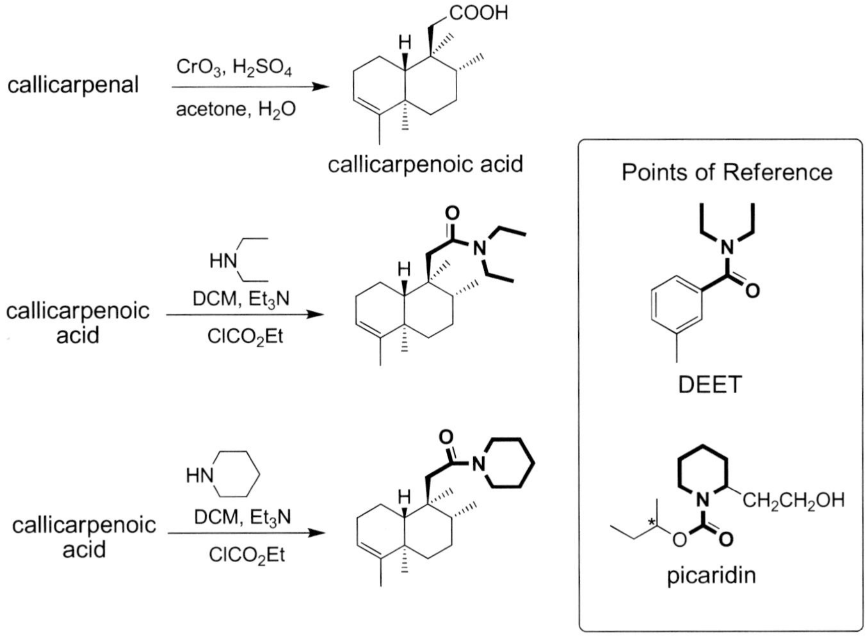

Figure 6. Synthetic conversion of callicarpenal to both its diethyl amine amide analog and piperidine amide analogs. DEET and picaridin shown as points of reference with similarities in bold.

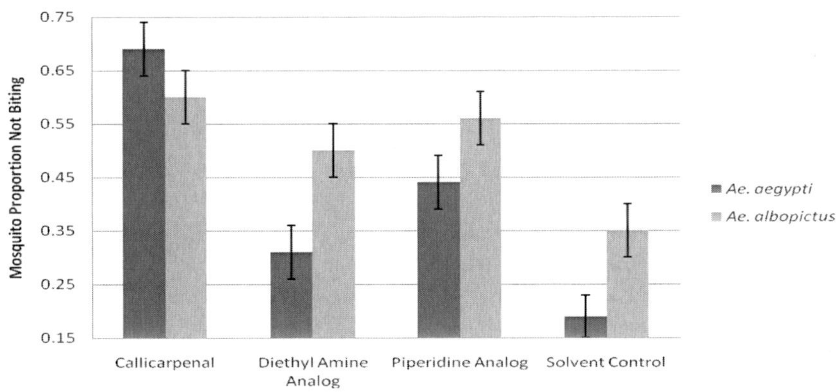

Figure 7. Callicarpenal and its diethyl amine and piperidine analogs at 25 nmoles/cm^2 versus Ae. aegypti and Ae. albopictus.

Callicarpenal and Intermedeol Extraction Methods

Due to commercial interest in developing callicarpenal as a natural product based insect repellent and the need for additional quantities for human toxicity studies, it is important to determine how we might efficiently obtain additional quantities of callicarpenal and intermedeol for such studies. One option is to obtain the compounds directly from the source plant via collection of the leaves, extraction and purification. The study described herein contrasts four extraction methods. Hexane is used as the extremely nonplolar solvent, methylene chloride (DCM) as the solvent with intermediate polarity, methanol as the most polar solvent and steam distillation as the last method.

All extraction methods utilized the same batch of *C. americana* which had been air-dried in a fume hood for 48 hours. For solvent extractions approximately 60 grams of plant material was extracted twice using 1L of the respective solvent at room temperature with subsequent drying, resuspension and analysis by GCMS in triplicate. For steam distillations, the exhaustive extraction method was described previously (*4*) and essential oil was resuspended and analyzed by GCMS in triplicate.

Quantitative analysis was performed using previously described GCMS analytical methods (*4*). A standard curve was generated for callicarpenal across a range from 1.02 to 0.001 mg/mL providing an R^2 of 0.999. Quantitation was performed by generating response factors for each analyte from the total ion chromatograms and applying this to the respective analytes of interest (Figure 8).

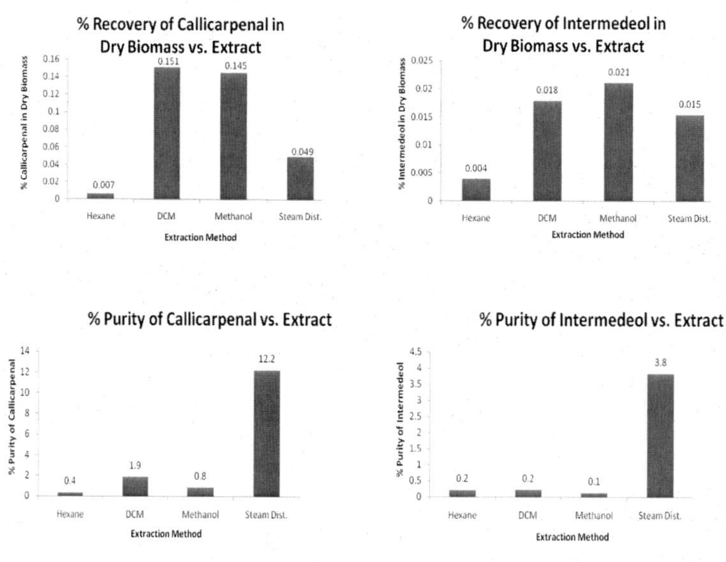

Figure 8. Recovery and purity analysis of callicarpenal and intermedeol in hexane, methylene chloride (DCM), methanol, and steam distillation extracts.

In the analysis of extracts for callicarpenal, both the DCM and methanol extraction methods resulted in the highest recovery of callicarpenal from dry biomass. Approximately 0.15% by weight of callicarpenal was present in dry biomass. Surprisingly, only 0.049% callicarpenal was recovered using an exhaustive steam distillation approach, or nearly one third that obtained from the DCM and methanol methods. For the analysis of intermedeol, the methanol extraction yielded the highest recovery with 0.02 % from dry biomass. This was followed closely by DCM and steam distillation with 0.018% and 0.015% respectively.

Purity analysis for each analyte in the various extracts proved to be extremely useful. The highest percent purity of callicarpenal (12.2%) was obtained in the essential oil produced via steam distillation. The next highest purity of 1.9% was obtained in the DCM extract. Similarly, the percent purity of intermedeol was also highest in the essential oil.

Callicarpenal Seasonal Variation

For reasons explained above, it also seemed relevant to determine if there is an optimal season for collecting leaves of *C. americana* where the concentration of callicarpenal would be highest. A small study was conducted to evaluate the variation in the dry weight concentration of callicarpenal throughout a single growing season. This particular study utilized two nearby collection locations in Lafayette County, Mississippi as sources for plant material throughout the season. A collection was performed at least once each month from the time leaves first appeared to the time their color began to change in the Fall throughout a single growing season. Briefly, 1.5 grams of air dried leaves were extracted with 200 mL of methanol in a Soxhlet extraction apparatus. Extraction solvent was removed and brought to volume in a 250 mL volumetric flask and directly injected and analyzed by GCMS as described above.

Figure 9 is a plot of seasonal variation of callicarpenal collected from two separate locations each month for one growing season. Both locations were in Lafayette County, Mississippi within Holy Spring National Forest. The percentage of callicarpenal present appears to peak during the months of May and June at approximately 0.045% (wt/wt) in dry leaves. The concentration continues to decline throughout the year until a mimimum is reached in the late summer months. The concentration in August is roughly half of its concentration at its peak in the Spring. Clearly the data leads us to believe that collection of leaves for the purpose of obtaining callicarpenal is optimium in the months of May and June.

Callicarpenal Sourcing and Scale-Up

One of the common problems encountered with developing a natural product into a commercial product or active ingredient is that of obtaining sufficient supplies at a reasonable cost to produce a product. This issue is dealt with routinely in the pharmaceutical industry where the production scale is much smaller than that typically found in the agrochemical industry. The high cost of pharmaceuticals can justify the cost associated with isolating and purifying natural products from their respective source. Unfortunately, the cost must be made much cheaper for agrochemicals to be economically viable. The situation is further complicated when dealing with compounds such as callicarpenal which contain four stereogenic centers making synthetic alternatives much more difficult. Scheme 1 depicts some of the alternatives for large scale production of callicarpenal, and each of these will be discussed below.

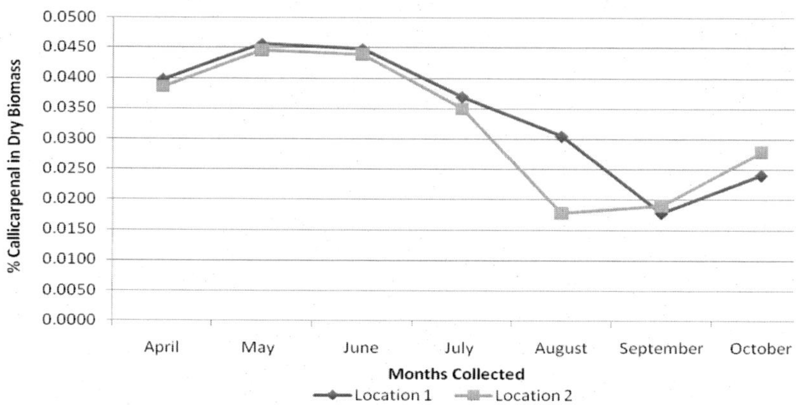

Figure 9. Seasonal variation of callicarpenal collected from two separate locations in Mississippi.

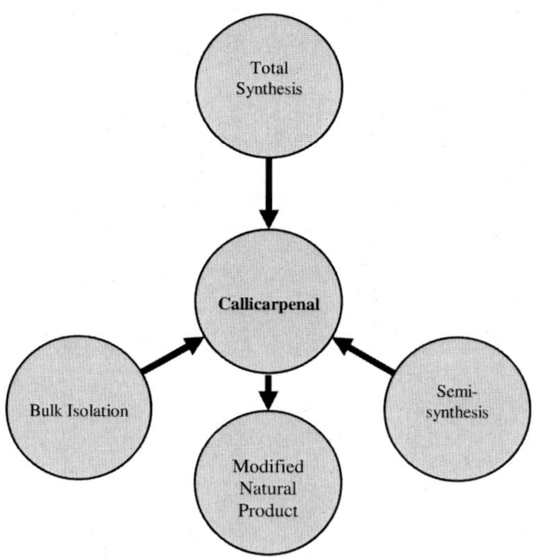

Scheme 1. Schematic representation of possible alternatives for producing sufficient quantities of callicarpenal for commercial parties.

Alternatives discussed in the above sections on isolation methods would suggest that callicarpenal could certainly be sourced directly from *C. americana* leaves by collecting during the months of May and June. Bulk purification from leaves would likely involve exhaustive steam distillation as the preferred method for obtaining callicarpenal primarily due to the higher puritiy of the extract when using this method. Despite the fact that methanol may yield a higher recovery of callicarpenal, the difficulties with purification could be problematic.

Total synthesis of callicarpenal is also a viable option and there are at least two reports in the literature on synthetic methods for producing callicarpenal (*13*, *14*). Both methods were not targeting callicarpenal as the final product, but instead it was produced as an intermediate in the total synthesis of more complex natural compounds. Hagiwara et al. (*13*) reported an enantioselective synthesis of callicarpenal in more than 20 steps, while Ling et. al. (*14*) synthesized callicarpenal in fewer than 15 steps. A semi-synthetic approach may be possible beginning from either a larger or smaller compound than callicarpenal; however, such procedures and/or methods are not in the literature. Such semi-synthetic methods are commonly used to produce pharmaceutical active ingredients such as vinblastline, vincristine, and docetaxel (*15*).

An even more desirable approach would be to better understand the relationship between structure and activity by completing a thorough structure–activity relationship (SAR) study on callicarpenal in hopes of such studies leading to more effective and cheaper to produce analog of callicarpenal. The preliminary SAR study completed by Cantrell et. al. (*10*) and described above is a first step in this direction.

Acknowledgments

This study was supported, in part, by a Deployed War-Fighter Protection Research Program Grant funded by the U.S. Department of Defense through the Armed Forces Pest Management Board.

References

1. First Growth in Global Agrochemical Market for a Decade; Agrow 466; February 18, 2005.
2. Nauen, R. *Pest Manage. Sci.* **2006**, *62*, 690–692.
3. p-Menthane-3,8-diol (011550) Fact Sheet. U. S. Environmental Protection Agency. www.epa.gov/pesticides/biopesticides/ingredients/factsheets/factsheet_011550.htm. (accessed June 23, 2005).
4. Cantrell, C. L.; Klun, J. A.; Bryson, C. T.; Kobaisy, M.; Duke, S. O. *J. Agric. Food Chem.* **2005**, *53*, 5948–5953.
5. Krajick, K. *Science* **2006**, *313*, 36–38.
6. Wedge, D. E.; Klun, J. A.; Tabanca, N.; Demirci, B.; Temel, O.; Husnu Can Baser, K.; Liu, Z.; Zhang, S.; Cantrell, C. L.; Zhang, J. *J. Agric. Food Chem.* **2009**, *57*, 464–470.

7. Cantrell, C. L.; Klun, J.; Duke, S. O. Novel Clerodanes and Methods for Repelling Arthropods. U.S. Patent application 2006/0235071 A1, 3/14/2006.
8. Carrol, J.; Cantrell, C. L.; Klun, J.; Kramer, M. *Exp. Appl. Acarol.* **2007**, *41*, 215–224.
9. Chen, J.; Cantrell, C. L.; Duke, S. O.; Allen, M. L. *J. Econ. Entomol.* **2008**, *101*, 265–271.
10. Soares, S. F.; Borges, L. M. F; Braga, R. S.; Ferreira, L. L.; Louly, C. C. B.; Tresvenzol, L. M. F.; Realino de Paula, J.; Ferri, P. H. *Vet. Parisitol.*, in press.
11. Soares, S. F. Thesis. Estudo da repelência de extratos de plantas e do DEET (n, n-diethyl15 m-toluamide) em *Amblyomma cajennense* (Acari: Ixodidae), 16 Universidade Federal de Goiás, Goiânia, 2008.
12. Cantrell, C. L.; Klun, J. A.; Pridgeon, J.; Becnel, J.; Green, S.; Fronczek, F. R. *Chem. Biodiversity* **2009**, *6*, 447–458.
13. Hagiwara, H.; Inome, K.; Uda, H. *J. Chem. Soc., Perkin Trans. 1* **1995**, *7*, 757–764.
14. Ling, T.; Poupon, E.; Rueden, E. J.; Kim, S. H.; Theodorakis, E. A. *J. Am. Chem. Soc.* **2002**, *124*, 12261–12267.
15. *Basic and Clinical Pharmacology*, 10th ed.; Katzung, B. G., Ed.; 2007, McGraw Hill Companies, Inc.

Chapter 4

Catnip Essential Oil and Its Nepetalactone Isomers as Repellents for Mosquitoes

Christopher J. Peterson[1] and Joel R. Coats[*,1]

[1]Department of Entomology, Iowa State University, Ames, Iowa 50011
[*]E-mail: jcoats@iastate.edu

The effect of catnip *Nepeta cataria* essential oil and two isomers of nepetalactone, the major components, on the distribution of *Aedes aegypti* (yellow fever mosquito) mosquitoes in a static-air olfactometer response was examined to determine their activity as spatial repellents. A glass cylinder was used as a choice-test chamber. The catnip (*Nepeta cataria*) essential oil, as well as the *E,Z-* and *Z,E*-isomers of nepetalactone were significantly repellent after application of one ml of 1% and 0.1% solution to filter paper (conc. of 157 and 15.7 μg/cm^2). Diethyl-*m*-toluamide (DEET) a positive control was significantly repellent at 157 μg/cm^2 in this assay. Both nepetalactone isomers and the catnip essential oil had excellent spatial repellency while DEET only exhibited spatial repellency at higher concentrations. The bioassay allowed for definition of and delineation between spatial and contact repellency.

Keywords: *Aedes aegypti*; repellent; *Nepeta cataria*; nepetalactone; spatial repellent

Introduction

For several decades, diethyl-m-toluamide (DEET) has been the most widely used insect repellent available. First synthesized in the early 1950s, DEET is usually regarded as safe, but up to 50% of the applied dose of DEET may be absorbed into the skin within six hours, and toxic effects have occasionally been documented in the literature, e.g. (*1*), (*2*), (*3*). In the United States, citronella is a popular botanical ingredient in insect repellent formulations. Candles and incense containing oil of citronella are also sold as insect repellents. It was reported (*4*) that citronella candles or incense were ineffective for reducing the biting pressure of mosquitoes. There is a strong desire among consumers for alternative choices for insect repellents, including high interest in natural alternatives.

The mint catnip *Nepeta cataria* L. is a common plant that frequently grows as a weed in many parts of the United States. Folklore had considered it an insect repellent for decades. Its oil was shown to keep ants from scavenging on a dead insect (*5*). Later it was shown to reduce the amount of time spent by German cockroaches *Blattella germanica* (L.) on the treated side of a choice-test arena (*6*, *7*). The essential oil of catnip has been shown to consist primarily of nepetalactone (70 to 98%), which is present primarily as two isomers, *Z,E*- and *E,Z*-nepetalactone (Figure 1) (*8*). Peterson (*7*) reports a separation process for these two isomers. The current study examines the effect of this essential oil and the individual isomers of nepetalactone on the distribution of yellow fever mosquitoes *Aedes aegypti* L. in a choice-test chamber. We used a 9 x 60-cm static-air glass repellency chamber to compare the effects of the catnip essential oil, the two pure individual isomers of nepetalactone at two concentrations (157 and 15.7 $\mu g/cm^2$) on the mosquitoes, and DEET at three concentrations (1572, 786, and 157 $\mu g/cm^2$).

Materials and Methods

Insects. A colony of *Aedes aegypti* was established in the summer of 1999 from wild mosquitoes collected in Costa Rica. The colony was blood-fed on a sedated rabbit. Eggs from mosquitoes were dried and stored in an incubator until needed. Eggs were placed in deoxygenated water and two to three drops of ground fish food were added to the water to feed the larvae. Pupae were removed from the larval pans as they appeared and were placed into mesh-covered paper cups. Adults were removed as they emerged by using an aspirator. The adults were separated by sex, and the females were retained for the repellency tests. The mosquitoes were allowed to feed on a cotton ball soaked with 10% (0.3 M) sucrose solution for four days before testing. The cotton ball was removed about 24 hr before the test was run.

Catnip Essential Oil and Nepetalactone isomers. Catnip plants were collected in the wild in Ames, Iowa. The catnip essential oil was obtained by steam distillation as described in Peterson (*9*). The nepetalactone from the catnip essential oil was identified by high performance liquid chromatography (HPLC) to consist of 85% *Z,E*-nepetalactone and 15% *E,Z*-nepetalactone. *Z,E*-nepetalactone was purchased as "catnip oil" from Kong Pet Products, Golden, CO. The HPLC method used an Hewlett-Packard 1100 HPLC with a Pirkle Covalent Phenylglycine Hi-chrom preparative column (25 cm x 10 mm I.D., 5μm S5NH Modified Shereosorb) and a mobile phase of 9:1 hexane:ethyl acetate at 2.5 ml/min flow rate, and detection with a Spectroflow 757 variable-wavelength UV-detector at 254 nm. Analysis of this oil determined that it consisted of 97.5% *Z,E*-nepetalactone, 0.8% *E,Z*-nepetalactone and 1.7% of an unknown. *E,Z*-Nepetalactone was purified from the catnip essential oil by preparative thin-layer chromatography as previously described (*10*). Diethyl-*m*-toluamide (DEET, 97%) was purchased from Aldrich Chemical Co., Milwaukee, WI.

Bioassay. A static-air choice-test apparatus consisted of a 9 x 60-cm section of glass tubing with a 2-cm hole drilled at the midpoint along the length for central introduction of the insects. One ml of test solution (oil, isomer, or DEET dissolved in acetone) was pipetted on a 9-cm filter paper and allowed to dry for five minutes. Another filter paper was treated with 1 ml certified acetone and allowed to dry for five minutes. The filter papers were placed inside the lids of 9-cm plastic petri dishes, and the lids were placed over the ends of the glass tube. The position of the treated side, to the right or to the left, was selected by using a random-number table. Twenty unmated adult female mosquitoes (four-day post emergence) were starved for 24 hours prior to the test, then anesthetized with carbon dioxide, and then introduced into the tube by using an aspirator. The insects were allowed to disperse through the tube for 15 minutes and the number of insects in each side (30 cm) were counted. A preliminary test was run with DEET at 1,572 and 786 $\mu g/cm^2$ to confirm if the assay was capable of detecting repellency. The catnip essential oil (CNEO) and two pure isomers were tested at two treatment concentrations, a high concentration and a low concentration, consisting of 157 $\mu g/cm^2$ and 15.7 $\mu g/cm^2$, respectively, which were prepared by pipetting one ml of 1% or 0.1% solution of repellent. Acetone controls (application of 1 ml of acetone, the carrier solvent) were conducted for each concentration. Five replications were conducted. The repellency observed is considered to be "spatial" repellency, since the vapors of the treatment are responsible for the movement of the mosquitoes to positions further away from the treated filter paper. Percentage repellency was calculated by subtracting the number of insects present on the treated side from the number on the untreated side, then dividing by the total number of insect in the chamber and the multiplying by 100 to convert the result to a percentage. The number of insects on each side of the tube was compared by using a paired t-test to determine if a treatment significantly altered insect distribution. Analysis of variance (ANOVA) was calculated by hand to determine significance due to treatments.

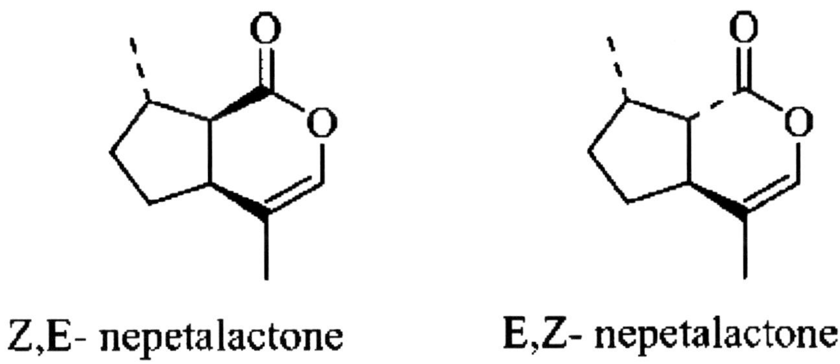

Figure 1. Z,E and E,Z racemic nepetalactone isomers in catnip.

Results

The results with 1,572 and 786 μg/cm^2 DEET show that the bioassay is capable of detecting spatial repellency for *A. aegypti* (Table 1). These levels were produced from 1 ml of 10% and 5% DEET, respectively; commercial DEET products have a range of concentrations of the active ingredient, from 7% to 50% and higher. Table 1 shows that at the relatively high concentrations (1,572 and 786 μg/cm^2, DEET was effective as a spatial repellent.

Table 1. Spatial repellency of a high concentration and a low concentration of catnip essential oil (CNEO), the isomes of nepetalactone, and positive control treatments to *Aedes aegypti*

Treatment	*% Repellency*[a]	*SEM*	*t-value*
	High conc. (157 μg/cm^2)		
CNEO	58.7	9.46	5.57*
Z,E isomer	49.6	14.8	3.62*
E,Z isomer	56.2	14.0	4.02*
DEET	10.0	7.41	1.38
Control	0.22	10.5	0.0
	Low conc. (15.7 μg/cm^2)		
CNEO	53.3	11.9	4.12*
Z,E isomer	45.6	12.4	3.67*
E,Z isomer	38.8	5.3	7.38*
DEET	9.7	6.76	1.4
Control	17.4	11.4	1.53
	Positive controls		
1,572 μg/cm^2 DEET	78.9	5.43	13.0*
786 μg/cm^2 DEET	85.3	3.94	9.90*
Control	-4.82	6.20	0.77

* Indicates that treatment was significantly different from a random distribution by paired t-test ($\alpha = 0.05$). [a] Percentage repellency = [(# of insects on untreated side - # insects on treated side)/total] x 100.

At 157 μg/cm2 the catnip essential oil (CNEO) and each isomer were all significantly (spatially) repellent, while the DEET did not show spatial repellency. At the lower concentration, from the application of 1 ml of 0.1% solution, the CNEO and the individual isomers were significantly repellent and the DEET was not. When DEET was evaluated, the distribution of the mosquitoes in the repellency chamber was not statistically different from the distribution for the "Control" treatment, in which no chemicals were applied other than the solvent.

Significant differences due to treatment were found for the high-concentration test ($F = 5.20$, $df = 4,20$), as well as the low-concentration test ($F = 3.48$, $df = 4,20$). These represent significance at the 99% and 95% levels, respectively. Significant differences in insect distribution were found by the paired t-test for CNEO, and for the *Z,E*- and *E,Z*-isomers. Significant effects on the distribution of the mosquitoes were not observed for DEET at either concentration, although is must be noted that the assay system was designed for testing spatial repellency. DEET is a very effective contact repellent; it is a more effective contact repellent than the monoterpenoids from catnip that were tested here.

Discussion

Thomas Eisner demonstrated that a drop of catnip essential oil repelled insects (*5*). Our first report on quantifying the repellency utilized a choice arena, with treated and untreated sides, with individual male German cockroaches (*6*). We first reported on the mosquito-repellent activity of the essential oil of catnip and of the two principal isomers of nepetalactone in 2001 (*9–11*). Quwenling and citronella are other monoterpenoid-based products that were marketed as mosquito repellents.

Essential oil of lemon eucalyptus was shown to be repellent, and its mostly active ingredient, p-menthane-3,8-diol, is also used as a natural, monoterpenoid insect repellent.

It was hoped that examination of the catnip essential oil and nepetalactone isomers would provide leads for development of mosquito-repellent products that may be safer and more accepted by the consumer than DEET. Other laboratories have also evaluated the oil of catnip and compared it to DEET. A triple-cage olfactometer was used to demonstrate that catnip oil was a better spatial repellent but poorer contact repellent, compared to DEET (*12*), while another study found it to be as effective as DEET in the K & D module *in vitro* assay, but not as good in bite-prevention on human subjects (*13*). Catnip essential oil was also demonstrated to be repellent to subterranean termites (*14*). Another study employed a slow-release formulation and showed excellent spatial repellency against stable flies (*15*). Two U.S. patents were issued (*16, 17*), and most recently DuPont has evaluated, patented and registered a synthetic dihydonepetalactone, which is a close analog of the natural monoterpenoid (*18*). Clearly the nepetalactone molecule has spurred new interest and efforts in development of natural, alternative repellents. A recent review of insect repellents has been published (*19*).

The simple static-air repellency chamber assay has provided a rapid and reliable tool for initial screening of potential repellent compounds, especially volatile ones that demonstrate spatial repellency. This assay is useful for comparison of spatial repellency activity of many essential oils and individual terpenoids. It does not involve any attractant or host, so the value is limited by that aspect, but it has proven to be useful for quantifying spatial, as well as contact, repellency; comparisons of the monoterpenoids in catnip essential oil with the sesquiterpenoids in osage orange essential oil showed that the monoterpenoids were more effective spatial repellents (in particular at early time points) and that the sesquiterpenoids were more effective contact repellents, although they slowly developed some spatial repellency as well (*20*), (*21*). A study of the ratio of *Z,E* to *E,Z* isomers produced by the catnip plants in Iowa revealed that the time of season clearly affected the ratio of the isomers (*22*). Blends of the catnip monoterpenoids with sesquiterpenoids yielded a repellent that demonstrated both early spatial repellency and long-lasting contact repellency (*23*).

Acknowledgments

We thank professors Lyric Bartholomay and Wayne Rowley of the Medical Entomology Laboratory at Iowa State University. The undergraduate laboratory assistants also assisted ably in the research: Leah Nemetz, Leah Jones, Gretchen Schultz, Sara Erickson, Mary Beth Penisten and Jaime Kutcher and the graduate student Jennifer Remmers. This is a journal paper of the Iowa Agriculture and Home Economics Experiment Station, Iowa State University, Ames, Iowa, project number 5075.

References

1. Roland, E. H.; Jan, J. E.; Rigg, J. M. Toxic encephalopathy in a child after brief exposure to insect repellents. *Can. Med. Assoc. J.* **1985**, *132*, 155–156.
2. Howard, I. M.; Howard, L. J. Contact urticaria syndrome (contact urticaria due to DEET). *Arch. Dermatol.* **1985**, *111*, 726–739.
3. Miller, J. D. Anaphylaxis associated with insect repellent (DEET). *New Engl. J. Med.* **1982**, *307*, 1342–1343.
4. Lindsay, L. R.; Surgeoner, GA.; Heal, J. D.; Gallivan, G. J. Evaluation of the efficacy of 3% citronella candles and 5% citronella incense for protection against field populations of *Aedes* mosquitoes. *J. Am. Mosq. Control Assoc.* **1996**, *12*, 293–294.
5. Eisner, T. Catnip: Its raison d'etre. *Science* **1964**, *146*, 1318–1320.
6. Peterson, C. J.; Nemetz, L. T.; Jones, L. M.; Coats, J. R. *Repellent Activity of Catnip and Osage Orange Fruit to the German Cockroach.* 218th ACS National Meeting, Agrochemicals Division, Poster No. 123, New Orleans, LA, August 22–26, 1999.
7. Peterson, C. J.; Nemetz, L. T.; Jones, L. M.; Coats, J. R. Behavioral activity of catnip (Lamiaceae) essential oil components to the German cockroach (Blattodea: Blattellidae). *J. Econ. Entomol.* **2002**, *95*, 377–380.

8. Bates, R. B.; Sigel, C. W. Terpenoids: cis-trans and trans-cis nepetalactones. *Experientia* **1963**, *14*, 564–565.
9. Peterson, Chris, Rowley, W., Coats, J. R. *Examination of Two Essential Oils As Mosquito Repellents*. 222nd ACS National Meeting, Agrochemicals Division, Poster No. 73, Chicago, IL, August 26–30, 2001.
10. Peterson, C. J. 2001. Insect Repellents of Natural Origin: Catnip and Osage Organge. Ph.D. Dissertation, Iowa State University, Ames, Iowa.
11. Peterson, C.; Coats, J. Insect repellents: Past, present and future. *Pestic. Outlook* **2001**, *12*, 154–158.
12. Bernier, U.; Furman, K. D.; Kline, D. L.; Allan, S. A.; Barnard, D. R. Comparison of contact and spatial repellency of catnip oil and N,N-diethyl-3-methylbenzamide (DEET) against mosquitoes. *J. Med. Entomol.* **2005**, *42*, 306–311.
13. Chauhan, K.; Klun, J.; Debboun, M.; Kramer, M. Feeding deterrent effects of catnip oil components compared with two synthetic amides against *Aedes aegypti*. *J. Med. Entomol.* **2005**, *42* (4), 643–646.
14. Peterson, C. J.; Ems-Wilson, J. Catnip essential oil as a barrier to subterranean termites (Isoptera :Rhinotermitidae) in the laboratory. *J. Econ. Entomol.* **2003**, *96*, 1275–1282.
15. Zhu, J.; Dunlap, C.; Behle, R.; Berkebile, D.; Wienhold, B. Repellency of a wax-based catnip-oil formulation against stable flies. *J. Agric. Food Chem.* **2010**, *58*, 12320–12326.
16. Coats, J. R.;et al., U.S. Patent 6524605B1, 2003.
17. Coats, J. R.;et al., U.S. Patent 6207705, 2001.
18. Feaster, J. E.; Scialdone, M. A.; Todd, R. G.; Gonzalez, Y. I.; Foster, J. P.; Hallahan, D. L. Dihydronepetalactones deter feeding activity by mosquitoes, stable flies, deer ticks. *J. Med. Entomol.* **2009**, *46* (4), 832–840.
19. Paluch, G. E.; Bartholomay, L. C.; Coats, J. R. Mosquito repellents: A review of chemical structure diversity and olfaction. *Pest Manage. Sci.* **2010**, *66*, 925–935.
20. Coats, J.; Schultz, G.; Peterson, C. *Botanical Products As Repellents against Mosquitoes and Cockroaches, AGRO-16*; 226th ACS National Meeting, NewYork, September 7–11, 2003.
21. Schultz, G. E.; Peterson, C. J.; Coats, J. R. Natural Insect Repellents: Activity against Mosquitoes and Cockroaches. In *Natural Products for Pest Management*; ACS Symposiumm Series 927; Rimando, A. M., Duke, S. O., Eds.; American Chemical Society: Washington D.C., 2006; Chapter 13, pp 168–181.
22. Schultz, G.; Simbro, E.; Belden, J.; Zhu, J.; Coats, J. R. Catnip, *Nepeta cataria* (Lamiales: Lamiaceae), a closer look: Seasonal occurrence of nepetalactone isomers and comparative repellency of three terpenoids to insects. *Environ. Entomol.* **2004**, *33* (6), 1562–1569.
23. Paluch, G. E.; Zhu, J.; Bartholomay, L. C.; Coats, J. R. Amyris and Siam-wood Essential Oils: Insect Activity of Sesquiterpenes. In *Pesticides in Household, Structural and Residential Pest Management*; ACS Symposium Series 1015, Peterson, C. J., Stout, D. M., II, Eds.; American Chemical Society: Washington, DC, 2009; Chapter 2, pp 5–18.

Chapter 5

Plant Essential Oils as Repellents and Deterrents to Agricultural Pests

Murray B. Isman* and Saber Miresmailli

Faculty of Land and Food Systems, University of British Columbia, Vancouver, British Columbia V6T1Z4, Canada
***E-mail: murray.isman@ubc.ca**

The concept of using non-lethal behavior-modifying substances for management of arthropod pests of agricultural crops has been long touted but as yet largely unexploited on a commercial scale. Numerous natural products have demonstrated repellent, antifeedant or oviposition deterrent activities in laboratory bioassays using pest insects, but consistent efficacy under field conditions has seldom been achieved. This results in part from unpredictable interspecific differences in responses to particular compounds, and in part from the ability of insects to habituate to deterrent compounds on repeated or continuous exposure. Plant essential oils represent a relatively new class of natural insecticides efficacious against a wide range of pests. While their neurotoxicity to insects and mites is widely recognized, there is strong anecdotal evidence that in some contexts efficacy could be attributed in part to their actions as behavior modifiers (i.e. as repellents or deterrents). Results of behavioral bioassays that explore the potential of certain essential oils and their constituents as repellents and deterrents to some other agricultural pests will be presented. In addition, changing composition of terpenoid emissions from essential oils over time will be discussed in relation to their biological effects.

For over 60 years insect pest management has relied extensively, and often exclusively, on the use of acutely toxic synthetic chemical insecticides. The dominant products used have evolved over this period, from the chlorinated hydrocarbons (e.g., DDT and the cyclodienes) to the organophosphates and carbamates (e.g., parathion and carbaryl), to the pyrethroids (e.g., permethrin) and most recently to the neonicotinoids (e.g., imidacloprid) (Thacker, J.R.M. *An Introduction to Arthropod Pest Control*; Cambridge University Press: Cambridge, U.K., 2002). These insecticides have been highly favored by growers owing to the wide assortment of products and formulations, their demonstrated efficacy and cost/benefit ratios in large scale agricultural practice, and their rapid, "curative" action when pest populations approach economic thresholds and threaten economic crop loss. But history tells us that reliance and overuse of these products has come with hidden or at least less obvious costs – toxicity to a wide array of non-target organisms, widespread soil and groundwater contamination, poisonings of pesticide applicators and farmworkers and concern for chronic health impacts in the general population as a consequence of pesticide residues in food (Benbrook, C.M. Pest *Management at the Crossroads*; Consumers Union: Yonkers, NY, 1996).

Concerns over environmental and human health impacts have led to increasingly restrictive regulation of chemical insecticides, particularly in North America, the European Union and Japan, and indeed many insecticide products that once dominated agricultural pest management have disappeared from the market (*3*). Recent pesticide use data from the State of California, a jurisdiction with among the most intensive pesticide use in the world, provides strong evidence of the trend toward reduced pesticide use, as shown in Table 1.

While the quantities of many important insecticides applied have declined, the overall quantities still used remain very large. Indeed, conventional insecticides, once touted to be displaced by biopesticides by 2000, continue to be the cornerstone of agricultural insect pest management.

The US EPA introduced the concept of "reduced risk" insecticides in 1992. Insecticides meeting the criteria of reduced risk (viz. less toxicity to nontarget organisms, reduced probability of groundwater contamination, reduced probability of resistance development) include microbials, botanicals, synthetic insect growth regulators, and the most recently developed conventional products that act through novel mechanisms-of-action. Given the popular and intuitive appeal of biopesticides (includes microorganisms, naturally occurring compounds and pheromones), it is interesting to note the 50% reduction in their use in California from 1998-2008, mirroring reductions for some of the traditional insecticides the biopesticides were expected to replace (Table 1). Moreover, the reduction in use of biopesticides over the decade is more than double that of all pesticides combined (-25.5%).

Table 1. Use of selected insecticides in California in 1998 and 2008

Insecticide	*Pounds applied 1998*	*Pounds applied 2008*	*% change*
Chlorpyrifos	2,451,980	1,350,399	-45.0
Diazinon	901,388	256, 218	-71.6
Malathion	663,200	474,863	-28.4
Methomyl	666,694	243,064	-63.5
Phosmet	645,380	339,696	-47.4
Biopesticides (all)	1,440,924	695,553	-51.3
Mineral and petroleum oils	27,254,084	25,716,688	-5.6
All pesticides	216,811,299	161,531,155	-25.5

SOURCE: California Department of Pesticide Regulation (*4*)

Plant Essential Oils as Insecticides

In spite of decades of research worldwide, only a handful of plant natural products have been successfully commercialized as botanical insecticides. Among those currently in the marketplace are pesticides based on plant essential oils. Facilitating commercialization of essential oil-based pesticides in the U.S. has been the exemption of certain oils from EPA registration, owing to their widespread use in foods, beverages and cosmetics. Well known examples are oils of rosemary, cinnamon, cloves, lemongrass, thyme and various mint species (*5*). That most of those are available as industrial commodities dramatically lessens the issues of supply and cost. Further advantages of these oils as insecticides are their relatively low acute mammalian toxicity (and therefore margin for human safety), their relatively short half-lives in the environment, and conversely, their rapid knockdown effect on many insects and related arthropods.

Chemically, plant essential oils are characterized by often complex mixtures of monoterpenes, sesquiterpenes and related phenols. It is these low molecular weight volatile compounds that account for the fragrances of the oils. In isolation, a number of these terpenoids have demonstrated contact and fumigant toxicity to insects; there is evidence that they also have sublethal repellent and/or deterrent effects in some insects (*5*). Many of these biological actions may be attributable to their action as agonists of octopamine (*6–8*), an arthropod neuromodulator, although there is emerging evidence that some oils or their constituents may have alternative targets in the insect nervous system (*9*). An interesting aspect of the toxicity of plant essential oils to insects and mites is the potential internal synergy of constituents – those deemed inactive in isolation have been demonstrated to boost the potency of those that are active in isolation (*10, 13*).

Owing to the useful bioactivity and relative availability of essential oils, together with their regulatory exemption in the US, insecticides based on certain oils have been commercialized for industrial, agricultural and consumer markets. While particularly effective for pest control in the home and garden, an agricultural insecticide based on rosemary and peppermint oils as active ingredients has proven successful against softbodied insects and mites in several crops. For example, a field trial for control of western flower thrips, *Franklinella occidentalis*, on strawberry in Salinas, California demonstrated that the essential oil-based product was more efficacious against thrips three weeks after treatment than spinosad, a fermentation-based insecticide prized by organic growers (*14*). Moreover, a tank mixture of the essential oil product and spinosad, each at one-half of their respective recommended label rates, was as effective as spinosad at the recommended rate. But the field efficacy observed raises an interesting question: does behavioral disruption – deterrence or repellence – contribute to overall efficacy?

Plant Essential Oils as Repellents and Deterrents

We had previously demonstrated in the laboratory that 1% rosemary oil can repel twospotted spider mites for at least 48 hours in a bean leaf choice test (*15*), and similarly that the oil reduced oviposition of greenhouse whitefly on tomato plants sprayed with 1% rosemary oil (*16*). Greenhouse experiments in Japan also demonstrated that rosemary oil could disrupt the settling behavior of green peach aphids on host plants ((*17*), also see (*5*)). Do these effects occur under field conditions? How persistent might the repellent effect be in the field if the oils evaporate rapidly from plant surfaces? We chose to explore the repellent/deterrent effect of plant essential oils under controlled laboratory conditions to better understand the dynamics of insect response over time to the oils. It is worth noting here that a repellent, by definition, causes oriented movement of an organism away from the source, whereas a deterrent simply prevents a behavior. In practice these actions can be difficult to distinguish which may explain why the terms of often used interchangeably. Note, for example, that we speak of "insect repellents", whereas many so-called products might actually work by disrupting alighting or probing behaviors of mosquitoes and other bloodfeeders, rather than causing them specifically to fly away from our skin. In addition to insect behaviors, we were also interested in the pattern of volatilization of an essential oil – do the major constituents all evaporate at the same rate, or does the composition of volatiles "evolve" over time?

To address these questions we conducted laboratory bioassays with two crop pests and one stored product pest. We tested the responses of adult twospotted spider mites, *Tetranychus urticae* (Acari: Tetranychidae) and confused flour beetles, *Tribolium confusum* (Coleoptera: Tenebrionidae) to rosemary oil, and those of 1st instar obliquebanded leafrollers, *Choristoneura rosaceana* (Lepidoptera: Tortricidae) to a commercial insect repellent (EcoSMART Insect Repellent™). All three species were maintained in continuous culture for at least three months in the absence of pesticides. Mites were reared on greenhouse-grown broad bean plants, beetles were reared on bran flakes, and leafrollers were reared on an artificial medium.

To analyze volatiles emanating from the oil and repellent we used a zNose™ – an ultrafast portable gas chromatograph. The zNose™ inlet, valve and initial column temperature were 200°C, 165°C, and 40°C, respectively. During analyses, the column temperature was increased at 10°C/sec to 200°C. The surface acoustic wave (SAW) sensor was kept at 50°C and the trap at 250°C. The helium flow during the 10 sec sampling period was set at 3.00 ccm. The sampling period was set for 10 sec at a sample flow of 20 ccm, after which the system switched to 20 sec of data acquisition. Thereafter, the sensor was heated to 150°C for 30 sec, and parameters (see above) were reset. The zNose system was tuned with an n-alkane solution and calibrated with neat reagents prior to its application in each experiment. As with other types of sensors, the response of the SAW sensor is proportional to the mass of the eluate, and calibration with the alkane standards permits quantitative analyses (see ref. (*11*) for a full discussion).

In the first experiment, we studied the effect of rosemary oil volatilization on the behavior of twospotted spider mites. The test arenas consisted of 12 glass plates (8 x 15 cm), ten treated with rosemary oil and two untreated as controls. Each plate was divided into 7 equally spaced regions and marked accordingly. For the treatment plates, the first region was painted with 20 µl of rosemary oil using a pipette. Every ten minutes, five adult mites were placed on the second (adjacent) region of the plate. Positions of these mites were then recorded for 10 minutes after their placement. The pattern of rosemary oil volatiles on each plate was recorded by placing the zNose over the first region of the plate immediately after placement of the mites. After 10 minutes mites were removed from the plate and replaced by naïve mites with no prior exposure to rosemary oil. The experiment ran for one hour, and was repeated twice.

Based on previous analysis of ten independent commercial samples of rosemary oil, the major constituents of rosemary oil are 1,8-cineole (52.1%), α-pinene (9.8%), camphor (9.0%), β-pinene (8.2%), camphene (4.9%) and *d*-limonene (3.8%)(figures in parentheses are averages) (*12*).

Our results indicated that overall, there was a significant effect of time on volatilization pattern of rosemary oil constituents (F [5, 44] = 101.984, $p < 0.05$; Wilk's lambda= 0.000)(Figure 1). Due to considerable variability, we did not find an overall statistically significant effect of compounds on the positions of mites on the test plates over the course of experiment. However, pair-wise comparisons of means showed that certain compounds had a significant effect on the position of mites at certain time intervals. As seen in Figure 1, the mites on the test plates moved away from the rosemary oil source at times matching peak release of 1,8-cineole, d-limonene and camphor.

In the second experiment, we studied the effect of rosemary oil on the behavior of adult flour beetles. The test arenas consisted of six 8x15 cm glass plates (5 treatments and 1 control). We covered the center of each treatment plate with a Petri dish (6 cm diameter) and applied 1 ml of rosemary oil evenly across the glass plate outside of the Petri dish and then removed the Petri dish and marked the untreated area. The zNose was then placed over the treated part of the plate. Five adult beetles were placed on the center of the plate. We measured the volatilization of rosemary oil constituents upon introduction of the beetles to the arena and every five minutes thereafter. Numbers of beetles that crossed into the treated zone were recorded every 5 minutes at which time beetles were replaced with five naïve beetles. The experiment ran for140 minutes after application or rosemary oil on the plates, and was repeated three times.

As in the first experiment, there was a significant effect of time on volatilization pattern of all compounds (F [27, 251] = 7603, $p < 0.05$; Wilk's lambda= 0.000) (Figure 2). Volatilization pattern of compounds also had a significant overall effect on the percentage of beetles crossing into the treated zone (F [6, 246] = 1.296, $p < 0.05$; Wilk's lambda= 0.969). Rosemary oil prevented beetles from crossing into the treated zone for 100 minutes, but 120 minutes after treatment, all naïve beetles "crossed the line" (Figure 2).

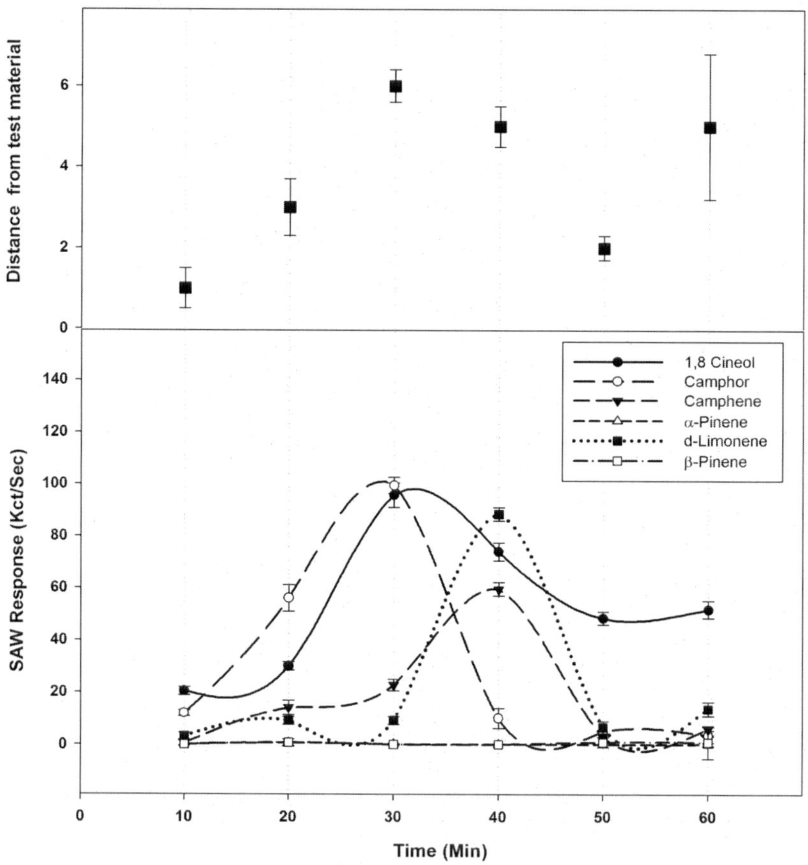

Figure 1. Comparing volatilization pattern of rosemary oil major constituents with position of naïve spider mites in a test arena over 60 minutes. Data points represent mean ± SD (n=20).

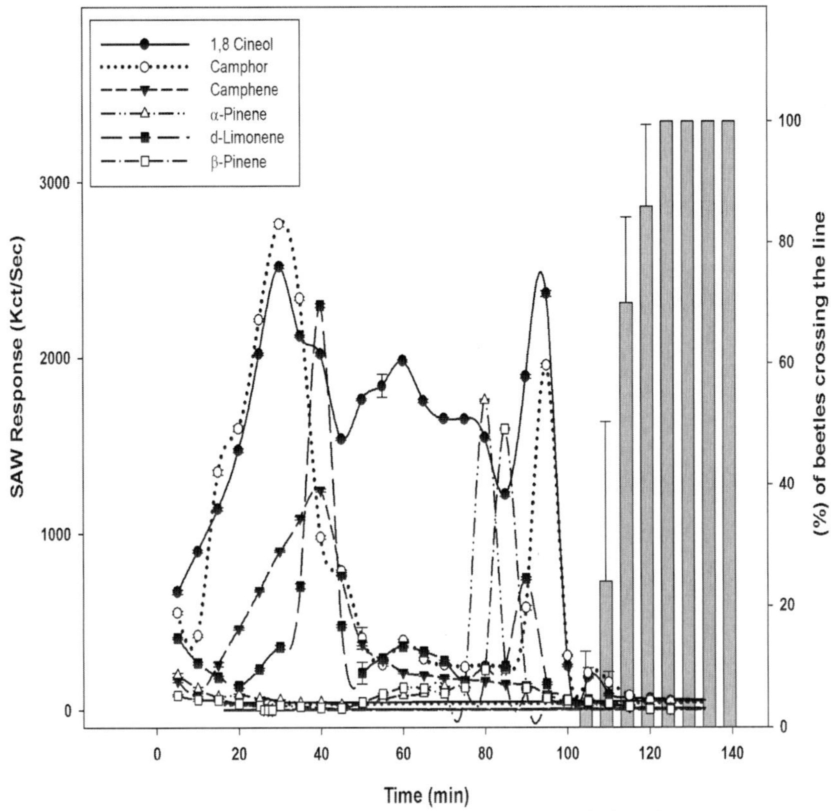

Figure 2. Comparing volatilization pattern of rosemary oil major constituents with percentage of naïve beetles crossing into the treated zone over time. Points and bars represent mean ± SD (n=15).

In the third experiment, we studied the effect of a commercial insect repellent on 1st instar obliquebanded leaf roller larvae. Six test plates were set up as in the second experiment. Treatment plates were treated with 1 ml of the commercial insect repellent. We installed a light source (40W clear light bulb) on one side of each plate to induce movement in the positively phototactic larva. Major constituents in this product include geraniol (0.6%), eugenol (0.4%), 1,8-cineole (0.2%) and citral (0.2%)(R. Bradbury, unpublished data).

Two larvae were placed in the center of the plate (untreated zone) at each data collection interval and allowed to move for five minutes. Numbers of larvae that crossed into the treated zone were recorded. Volatiles were analyzed and naive larvae added every five minutes for a total of 140 minutes.

Again there was a significant effect of time on the volatilization pattern of the repellent compounds (F [26, 107] = 263.914, $p < 0.05$; Wilk's lambda= 0.000). The volatilization pattern of compounds did not have a statistically significant overall effect on the percentage of larvae crossing the treatment line. Larvae started to cross the treatment line after 110 minutes; all larvae crossed into the treated zone after 130 minutes (Figure 3).

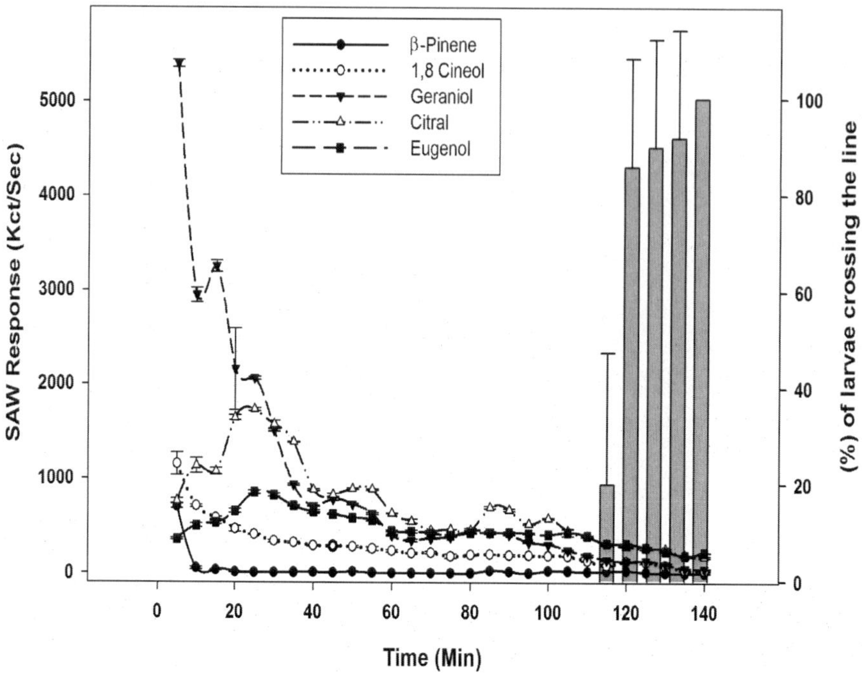

Figure 3. Comparing volatilization pattern of insect repellent major constituents with percentage of naïve larvae crossing the treatment line over time. Points and bars represent mean ± SD (n=5).

Previously, we showed that presence or absence of certain constituents in a mixture could significantly effect the efficacy of essential oil-based botanicals as contact pesticides (*10*, *12*). Results of the present study provide new insight in the behavioral impacts of these products as insect and mite repellents. Should these essential oils be used as repellents, efficacy could be tied to the volatilization pattern of constituents. Our study shows that the volatilization pattern of essential oils changes over time. In the case of rosemary oil (Figures 1 and 2), some constituents will only be present in the headspace after the level of other constituents decreases (i.e., *d*-limonene and camphene versus 1,8-cineol and camphor). These interactions can affect the behavior of insects and mites.

Our first experiment confirms this: mites were repelled the most between 20 and 40 minutes when most of the major constituents were present at maximum levels. In the second and third experiment, we demonstrated that there is a threshold for repellence, which is closely related to the volatilization level of constituents. The compounds might be still present as a residue of the mixture on the surface where it was applied, suggesting a deterrent effect rather than a repellent one. However, we did repeatedly observe leafrollers and beetles approach the treated zone and turn away without touching it, suggesting a true repellent effect albeit at close range. This repellence decreased when the volatilization levels decreased.

With the commercial insect repellent, we observed a gradual change in the level of volatiles, while for rosemary oil, volatiles decreased suddenly after 100 minutes. Although the reason for this phenomenon is not immediately clear, it might be possible to develop a particular mixture and/or formulate it such that it maintains levels of volatiles that decay gradually over longer periods of time, therefore, remaining more biologically effective. Microencapsulation or nanoparticle technology may be useful techniques for achieving prolonged release of volatiles, as has been demonstrated with garlic oil (*18*).

Insect Antifeedants (and Oviposition Deterrents) as Crop Protectants

Voluminous scientific literature indicates that terrestrial plants are a rich source of natural products that deter feeding of one or more species of insect under laboratory conditions. These observations should not be surprising: the *modus operandi* of plant chemical defense is primarily based on discouraging herbivory, rather than killing insects outright. This may explain why so few botanical insecticides have been commercialized – most plant natural products are not acutely toxic to insects, at least at doses comparable to those for synthetic insecticides (*19*). The concept of using insect antifeedants as nontoxic crop protectants is intuitively appealing and has been the subject of considerable research (*20*). While there have been a handful of studies reporting moderate efficacy of antifeedants under field conditions, not a single antifeedant has been commercialized for crop protection to date. Why is there such a big disconnect between laboratory science and commercial application in this regard?

The answer may lie in the plasticity of insect sensory physiology and feeding behavior. In the vast majority of studies reporting antifeedant effects of plant compounds, naïve insects are utilized and then discarded; seldom are the same insects tested twice. In so doing, many investigators have failed to observe an important phenomenon of insect behavior, and one that bears on the potential efficacy of antifeedants as crop protectants – habituation. Almost 15 years ago we demonstrated that the Asian armyworm *Spodotera litura* habituated to the extremely potent antifeedant azadirachtin, when challenged with azadirachtin-treated leaf discs on successive days (*21*). By the third day, feeding deterrence had dropped to 10%, compared to 60% for naïve larvae on the first day of testing. Moreover, when we exposed larvae to azadirachtin continuously, they habituated almost completely after 4.5 hours of feeding (*22*). That this phenomenon was not restricted to azadirachtin was later shown in our studies with the cabbage looper *Trichoplusia ni*, which habituated fully to several known antifeedants, both terpenoid and phenolic, and to complex plant extracts. Moreover, we even found that exposing loopers to certain feeding deterrents in the last larval stage negated the oviposition-deterring properties of those compounds to the subsequent moths (*23*). In other words, the memory of the compounds to which the larvae habituated was retained through metamorphasis from larva to pupa and from pupa to adult. In some cases, what is normally an oviposition deterrent to naïve moths became an oviposition stimulant for the "experienced" moths. This is all the more remarkable when considering that larvae and adult do not share gustatory organs; larval gustatory sensilla are found exclusively on the mouthparts, whereas adult lepidopterans "taste" plants using chemosensilla on their tarsi (feet) and on the ovipositor (egg-laying organ) itself.

What this means in practical terms is that an insect in constant contact with a deterrent applied to a crop may quickly overcome the deterrence and feed with impunity, unless of course the antifeedant also has some deleterious physiological effect on the insect, as in the case of azadirachtin. While azadirachtin is a potent antifeedant to many (but not all) insects, it is likely its insect growth regulatory activity that accounts for the efficacy of neem-based insecticides in commercial use (*3*). Habituation to sex attractant pheromones has been observed, and in some situations could be a factor limiting the success of mating disruption programs based on mass release of synthetic pheromone (*24*). Little is currently known about the potential for habituation in insects to repellent volatiles, but it would be logical to assume that this could pose a limitation to the use of repellents as crop protectants.

One aspect of synthetic insecticides that has greatly facilitated their acceptance by growers is their consistent efficacy in most insect management situations and contexts. The user need not have a detailed understanding of insect physiology or toxicology, and there is often a wide window with respect to timing of applications. In contrast, the use of insect behavior-modifying chemicals for crop protection requires a sound understanding of insect behavior and ecology. Efficacy for products of this type may depend to a far greater extent on such factors as timing, temperature and the formulation of the active ingredients compared to that of conventional insecticides. Our understanding of how and when repellents or deterrents could be effective for crop protection is very thin at the moment. In the case of the essential oil-based insecticides, it is entirely possible that much of the efficacy already seen in the field against certain pests may be a consequence of repellence rather than toxicity. Thus continued research is warranted, including observational studies in the field and further development in the laboratory of potential alternative crop protection products. The decline in the use of conventional insecticides (Table 1) provides an obvious impetus for such research.

Acknowledgments

Based on a symposium paper presented at the 238th ACS Annual Meeting in Washington, DC, August 16, 2009. We thank Nancy Brard for technical assistance, Pacific AgResearch for permission to use their unpublished data and Cristina Machial for comments on the manuscript. Supported by grants from MITACS, NSERC, Ecosafe Natural Products Inc. and EcoSMART Technologies Inc.

References

1. Thacker, J. R. M. *An Introduction to Arthropod Pest Control*; Cambridge University Press: Cambridge, U.K., 2002.
2. Benbrook, C. M. *Pest Management at the Crossroads*; Consumers Union: Yonkers, NY, 1996.
3. Isman, M. B. *Annu. Rev. Entomol.* **2006**, *51*, 45–66.
4. California Department of Pesticide Regulation. Summary of Pesticide Use Report Data 2008. http://www.cdpr.ca.gov/docs/pur/pur08rep/chmrpt08.pdf (accessed 20 August 2010).
5. Isman, M. B. *Crop Protect.* **2000**, *19*, 603–608.
6. Enan, E. *Comp. Biochem. Physiol. C* **2001**, *130*, 325–337.
7. Enan, E. *Arch. Insect Biochem. Physiol.* **2005**, *59*, 161–171.
8. Kostyukovsky, M.; Rafaeli, A.; Gileadi, C.; Demchenko, N.; Shaaya, E. *Pest Manage. Sci.* **2002**, *58*, 1101–1106.
9. Priestley, C. M.; Williamson, E. M.; Wafford, K. A.; Sattelle, D. B. *Br. J. Pharmacol.* **2003**, *140*, 1363–1372.
10. Miresmailli, S.; Bradbury, R.; Isman, M. B. *Pest Manage. Sci.* **2006**, *62*, 366–371.
11. Miresmailli, S.; Bradbury, R.; Isman, M. B. *Arthropod−Plant Interact.* **2010**, in press.
12. Isman, M. B.; Wilson, J. A.; Bradbury, R. *Pharm. Biol.* **2008**, *46*, 82–97.
13. Jiang, Z.; Akhtar, Y.; Bradbury, R.; Zhang, X.; Isman, M. B. *J. Agric. Food Chem.* **2009**, *57*, 4833–4837.
14. Pacific Ag Research, Salinas, CA, 2009, unpublished data.
15. Miresmailli, S.; Isman, M. B. *J. Econ. Entomol.* **2006**, *99*, 2015–2023.
16. Isman, M. B.; Miresmailli, S.; Machial, C. *Phytochem. Rev.* 2010, in press.
17. Hori, M. *J. Chem. Ecol.* **1998**, *24*, 1425–1432.
18. Yang, F. L.; Li, X. G.; Zhu, F.; Lei, C. L. *J. Agric. Food Chem.* **2009**, *57*, 10156–10162.
19. Isman, M. B. In *Biopesticides of Plant Origin*; Regnault-Roger, C., Philogene, B. J. R., Vincent, C., Eds.; Lavoisier: Paris, 2005; pp 283−291.
20. Isman, M. B. *Pestic. Outlook* **2002**, *13*, 152–157.
21. Bomford, M. K.; Isman, M. B. *Entomol. Exp. Appl.* **1996**, *81*, 307–313.
22. Akhtar, Y.; Rankin, C. H.; Isman, M. B. *J. Insect Behav.* **2004**, *16*, 811–831.
23. Akhtar, Y.; Isman, M. B. *J. Chem. Ecol.* **2003**, *29*, 1853–1870.
24. Witzgall, P.; Stelinski, L.; Gut, L.; Thomson, D. *Annu. Rev. Entomol.* **2008**, *53*, 503–522.

Chapter 6

Contact and Spatial Repellency from Catnip Essential Oil, *Nepeta cataria*, against Stable Fly, *Stomoxys calcitrans*, and Other Filth Flies

Junwei Jerry Zhu*

Agroecosystem Management Research Unit (AMRU), Agricultural Research Service (ARS), U.S. Department of Agriculture, 305 Entomology Hall, University of Nebraska, Lincoln, Nebraska 68583
***E-mail: Jerry.Zhu@ars.usda.gov**

A newly discovered botanical repellent, catnip oil (*Nepeta cataria*, L.), which includes its efficacy on feeding repellency, ovipositional deterrency and spatial repellency against stable fly, is described. It also discusses its practical applications, with the developed oil- and water-based formulation of catnip essential oil, for repelling biting flies on cattle under the field conditions. Finally it touches on the topic of the safety of using this product with presented toxicity data, and comparisons with other blood-sucking insect repellents.

Introduction

Filth flies are flies that develop in rotting, decaying or fermenting organic materials. There are four major filth fly pests of livestock, horn flies, face flies, house flies, and stable flies. In the United States, stable flies are the most important pests of livestock, especially on cattle. The painful bites and the behavioral and physiological responses they invoke in cattle cost US producers more than $2 billion per year (*1*). Stable flies are also capable of transmitting a large variety of pathogens including helminths, protozoans, bacteria, and viruses, some of which are primary agents of mortality in cattle (*2–4*). Furthermore, very few options currently exist for reducing the damage of stable flies on pastured cattle where they cause an estimated $1.3 billion in losses per year.

Stable flies utilize fermenting or decomposing vegetation as their breeding sites (*5–7*). The areas along the soil-to-concrete interface of the feed apron in the feedlot pens can generate about 80% of fly immatures at confined cattle facilities (*8*). Current practices for managing stable flies are limited to costly sanitation techniques and unsustainable insecticide applications, but are limited on confined animals (*9*, *10*). Zumpt (*2*) suggested that spraying cattle with repellents or applying contact insecticides to fly resting areas could suppress the build-up of fly densities significantly.

Plant derivatives, or botanical insecticides and repellents have been used against arthropods for at least two millennia in ancient China, Egypt, and India (*11*, *12*). In Europe and North America, the practice of using botanicals dates back more than 150 years (*13*). Recent studies have confirmed the effectiveness of repellent properties of plant essential oils against Dipteran blood-sucking insects, particularly in mosquitoes (*14–17*). Zhu et al. (*18*) reported that catnip (*Nepeta cataria L.*) essential oil acts as an extremely effective antifeedant/repellent against filth fly species, including stable flies, horn flies, face flies and house flies, in laboratory antifeedant bioassays. They have further demonstrated that catnip oil is a relatively safe repellent with an extremely low toxicity in rabbits and rats (*18*). However, the relatively short longevity may limit its practical application. Therefore, the development of a slow-released, but an effective formulation may prove its usefulness under field conditions. The lone use of repellents against biting flies in livestock animals may seem un-realistic, but a filth fly integrated management program involving a Push-Pull strategy can be successful. Similar strategies have been developed successfully for other agricultural and urban pests (*19*). Furthermore, the new findings of oviposition deterrence from catnip oil can be involved in the integrated filth fly management strategies to enhance their effectiveness, such as application of slow-released oviposition deterrent formulations around their potential breeding sites.

Stable Fly Olfactory and Gustatory Sensilla

Sensory organs on the antennae of insects are known to be used in locating mates, hosts, habitats, and oviposition sites (*20–22*). Studies of the antennal sensilla in Dipteran species have revealed an abundance of basiconic, coeloconic, and trichoid sensilla (*23–28*). In muscoid flies, most sensory organs used for the perception of chemical odorants are located on the funicle of antennae (*29–31*). These sensory organs have been reported to respond to various stimuli such as warmth, humidity, chemical odors, including ammonia, and carbon dioxide (*32–35*). Using scanning electron microscopic (SEM) and transmission electron microscopic (TEM) images, Tangtrakulwanich et al. (*36*) and Lewis (*37*) have described the details of external and internal structures of four major types of sensilla from the antenna of stable flies, which are classified as: 1) basiconic sensilla, 2) trichoid sensilla with three subtypes, 3) clavate sensilla, and 4) coeloconic sensilla (Figure 1). Among them, the distinctive pore structures have only been found on surfaces of basiconic and clavate sensilla, which suggest their olfactory functions. Volatile semiochemicals emitted from the host animals and their environments play major roles in mediating host location and oviposition site selection. Several odorant compounds from cow rumen, urine and manure have been identified that are attractive to stable flies (*38*, *39*). Gravid stable fly females are *capable* of selecting oviposition sites based on microbe-derived stimuli that indicate suitability of the substrate for larval development ((*40*); *unpublished observation*).

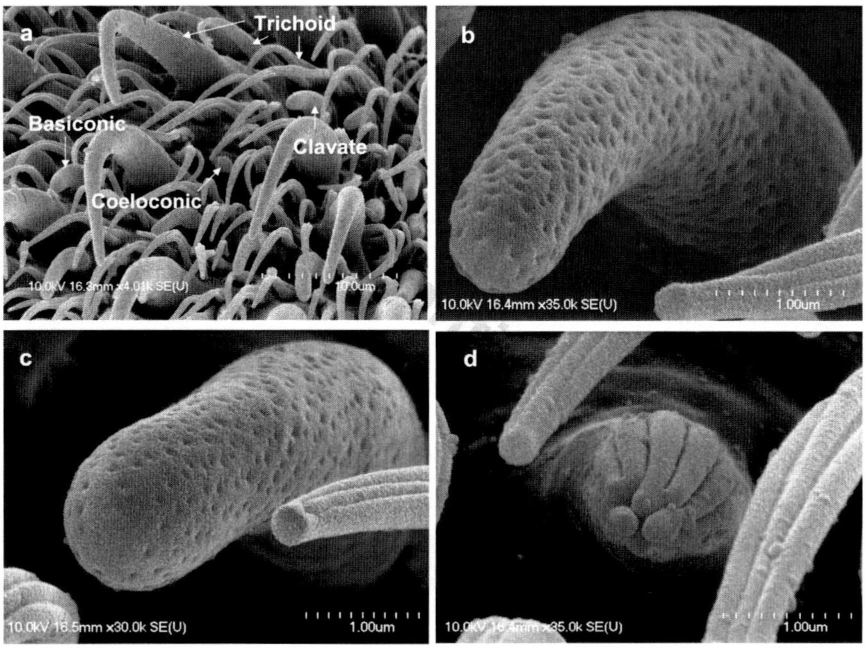

Figure 1. SEM micrographs of stable fly olfactory sensilla on antennae: (a) Dorsal view of the funicle showing the distribution of all sensilla types; (b-d) detailed views show shape differences between basiconic sensilla and clavate sensilla, with pore-structures on the wall surface, and the finger-like coeloconic sensillum.

Much of what is known concerning fly proboscis structure is due to the fact that some flies 'bite'. For stable flies, Stephens and Newstead (*41*) have described and illustrated the prestomal teeth with their lateral and terminal serrations, and the presence of petiolate blades. They have hypothesized that stable fly uses the concept of a carpenter's augar as a mode of beal penetration. Elzinga and Broce (*42*) have reported that the stable fly proboscis is extended with their everting labellar lobes that enable the exposed pseudotracheae to transfer the liquid directly to a central cavity (where a labial-gutter-epipharyngeal tube is located). We have discovered 6-8 pairs of contact-chemoreceptor sensilla on their labellum (Figure 2). The tip pore may be used to detect feeding stimulants/deterrents, as constant probing is observed, but no blood is observed to be in-taken, when feeding deterrents are present.

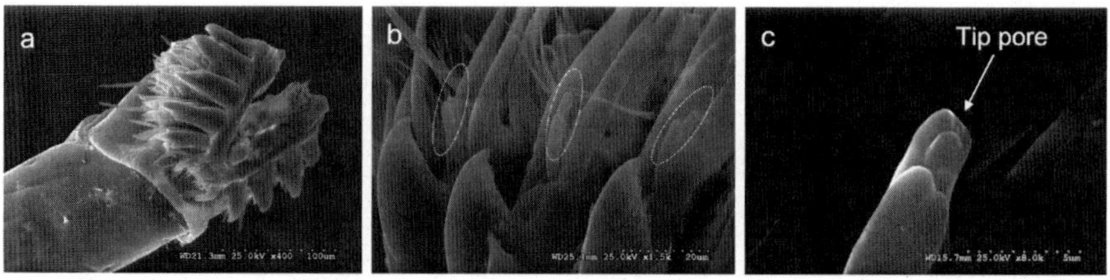

Figure 2. SEM micrographs of the stable fly (a) everted labellum (terminal view); (b) 3 pairs of chemo-contact sensilla; (c) tip pore of an individual sensillum.

Catnip Essential Oil as a Repellent against Arthropods

Many plant extracts have been identified as having repellent effects (*13*, *43*). Several articles have reported varying degrees of repellency of plant oils against blood-sucking insects, which include clover, peppermint, geranium, neem, and turmeric, etc. (*13*, *44–47*). Catnip (*Nepeta cataria*) is an herbaceous mint native to Eurasia and North Africa, and is also found in most of North America. It is well-known for its pseudo-narcotic effects on cats. During the last 10 years, it has been reported that topical application of catnip essential oil can effectively prevent biting by several disease-transmitting mosquito species, with additional evidence of spatial repellency (*17*, *48–50*). The chemical composition has been determined by gas chromatography-mass spectrometry (GC-MS) analysis, which shows 90% of *ZE*- and *EZ*-nepetalactone, and 10% of caryophyllene (*18*). Eisner (*51*) has reported that catnip oil repels up to 13 families of insects. Zhu et al. (*18*, *52*) have discovered that catnip oil shows strong repellency against several filth flies.

Development of Biting Fly Contact Repellent Assay

The laboratory bioassay using a six-well feeding reservoir system (K & D module) has been widely used for testing repellent efficacy on mosquitoes (*53*). Almost no study has been reported to develop a similar laboratory bioassay to screen potential repellent candidates on biting flies, without the burden of using animals. The K & D module was originally designed for testing mosquito repellents *in vitro*, however, during testing on biting flies, no flies were observed to successfully feed on the blood through the adapted membranes (Baudruche membrane or Edical collagen) above the feeding wells. Starved flies were observed probing the membrane aggressively during the testing period, but were incapable of penetrating the layer of membranes employed in the K&D module. By using the outer cover layer of the feminine napkin, Zhu and his colleagues (*18*) have successfully modified K&D module to adapt it to the mouth parts of stable flies, which enables them to cut through the layers for blood feeding (Figure 3). The control flies showed a feeding rate of 96-100% from hundreds of flies tested. Development of such an *in vitro* bioassay has contributed significantly for discovering novel repellents against various biting flies in future studies.

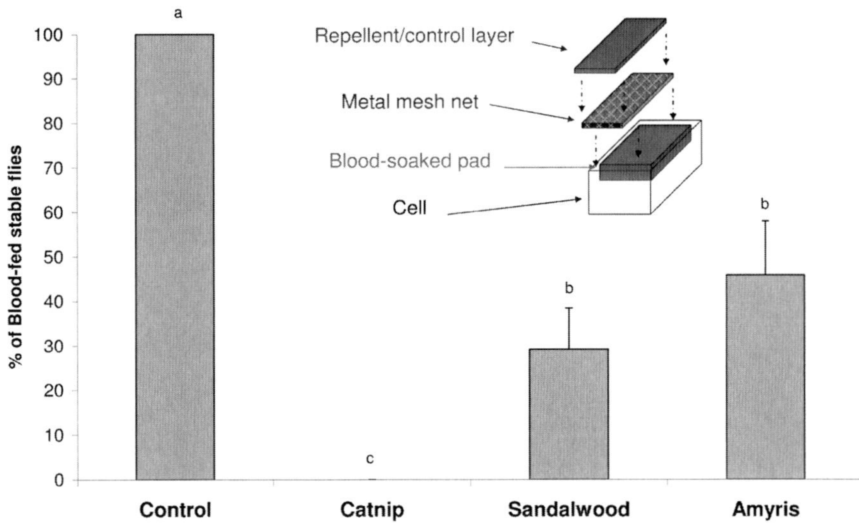

Figure 3. Percentages of feeding observed from starved stable flies treated with three different repellent oils in modified K&D modules (insert). Means with different letters are significantly different at P < 0.05 level (SAS version 9.1, performed on the Least-Square Means).

Feeding Repellency of Catnip Oil against Stable Flies

Catnip oil strongly repelled stable flies from blood feeding with a >98% repellency rate and was significantly higher than the other plant essential oils tested at a 20-mg dosage (Figure 3). The major constitutional components of catnip oil, ZE- and EZ-nepetalactone prevented stable flies from blood-feeding as effectively as catnip essential oil (Figure 4A). A significantly lower repellency (< 20%) was observed from caryophyllene. A further comparison test on repellency of catnip oil was carried out with several recently-identified deterrents and repellents against blood-feeding arthropods including (-)-isolongifolenone (J4-118), 2-methylpiperidinyl-3-cyclohexen-1-carboxamide (AI3-37220) and (1S,2'S)-2-methylpiperidinyl-3-cyclohexen-1-carboxamide (SS220), as well as the mostly commonly used mosquito repellent, N,N-diethyl-3-methylbenzamide (DEET) (*54–56*). It has been shown that the effectiveness of repellency from catnip oil and AI3-37220 is significantly higher than DEET (Figure 4B). No differences were observed among DEET, J4-118 and SS220 in preventing stable fly feeding.

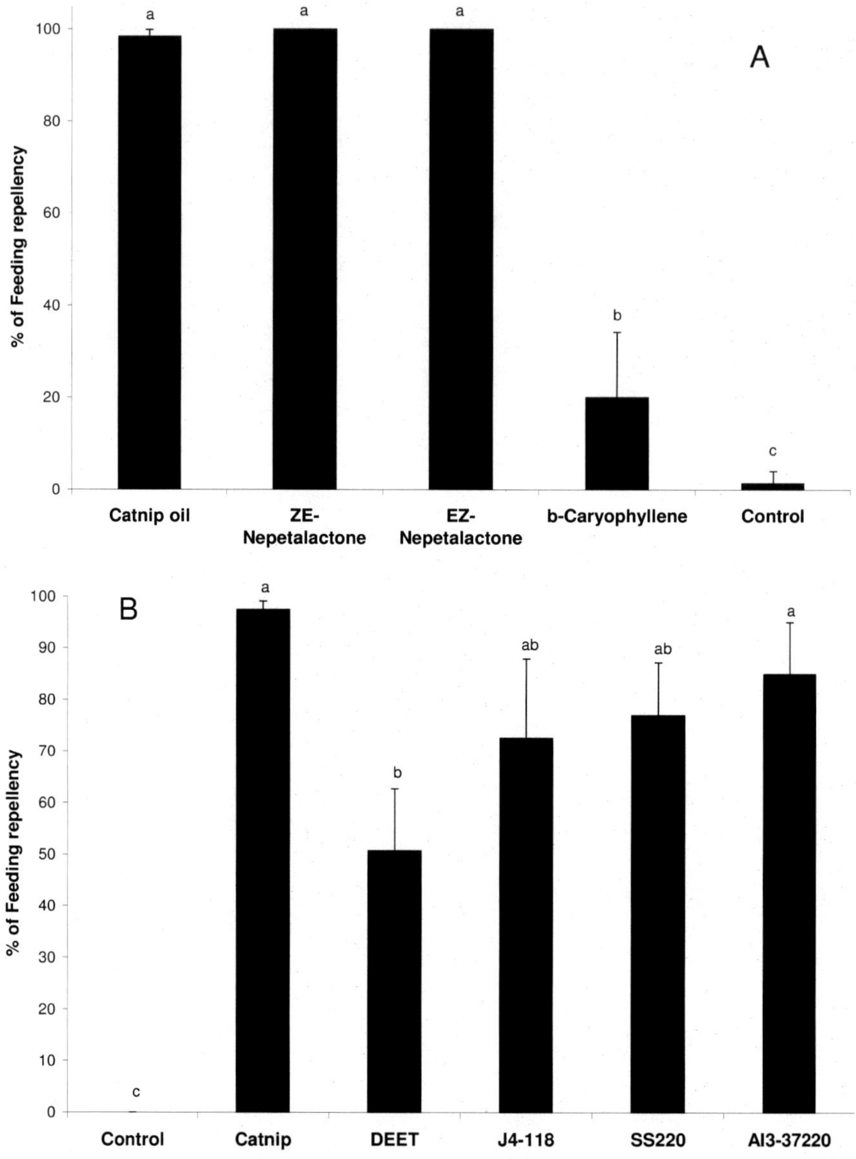

Figure 4. Mean Percentages of feeding repellency observed from starved stable flies treated with 20 mg of catnip oil and its compositional compounds (A), and to catnip oil and other recently-identified insect repellents at the same concentration (B) in a laboratory in vitro system. Means with different letters are significantly different at (P < 0.05, ANOVA followed by Duncan's test.

Spatial Repellency of Catnip Oil on Stable Flies and Their Olfactory Responses

While conducting catnip antifeedancy studies in modified K&D modules, Zhu et al. (*18*) also observed that tested stable flies in the catnip oil-treated cells tried to fly away from the repellent-treated surface. This indicated that catnip oil may also have a spatial repellency against stable flies. They have further designed one dispersal study to test their hypothesis, and demonstrated that catnip oil does significantly repel stable flies from the catnip oil-treated areas (Figure 5). Percentages of repellency ranged from 18% to 50% observed from the dispersal study during the 4-hour experimental period (Figure 5, bars). The analyses of accumulative, atmospheric concentrations of two nepetalactones absorbed by SPME fibers in the catnip oil-treated areas revealed a 6-fold increase of catnip atmospheric concentration 4 hours after initial exposure (Figure 5, line). These results suggest that the atmospheric concentration of catnip oil contributed significantly to the spatial repellency.

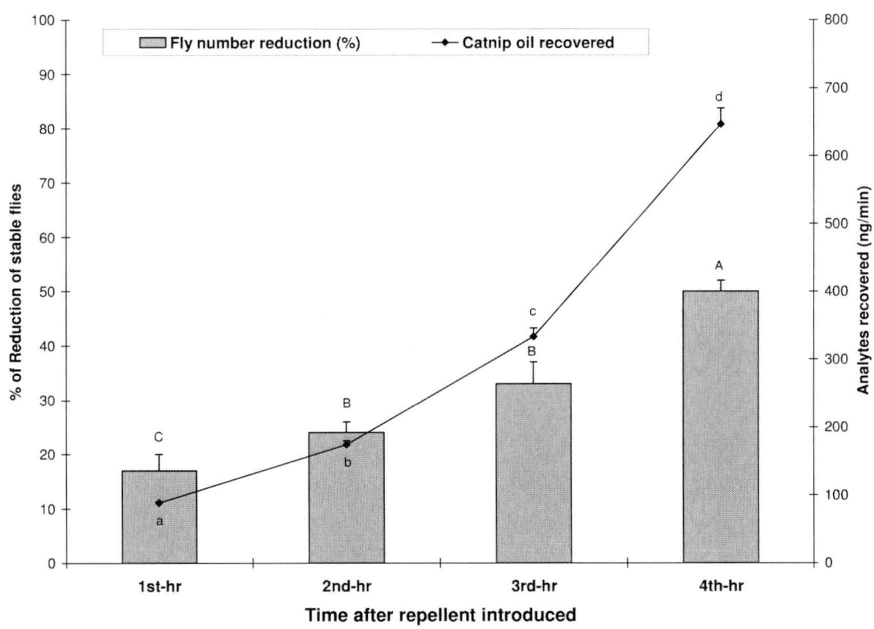

Figure 5. Reduction of stable fly numbers after catnip oil application at 4 μg/cm² concentration to filter paper in the screen cage (bars) and catnip oil volatiles recovered by solid phase extraction(five minute atmospheric sampling). Different letters on top of bars and lower case letters on the line are significantly different at the level of P < 0.05 according to ANOVA, separated by Scheffe test.

Bernier et al. (*50*) demonstrated that catnip oil acts as a spatial repellent against female *A. aegypti* mosquitoes in an olfactometer assay. A single cage olfactometer study on stable flies (*52*) has further shown that over 70% of flies were repelled from the catnip oil-treated port, compared with the control (Table 1). Stable flies were observed to be highly attracted to 1-octen-3-ol (a ruminant odorant found from animal breath) with an observed 75% of attractancy in the olfactometer study, but the attractiveness was reduced significantly when catnip oil was added (*39*, *57*, *58*).

Olfactory responses (electroantennogram) have been measured from stable fly antennae from a range of attractants from host animals and odorants associated with oviposition sites (*39*). It is possible that the same olfactory sensilla on stable fly antennae are also capable of detecting active repellent compounds, e.g. nepetalactones of catnip oil (Figure 6A). The EAG tests have further shown that DEET also elicits antennal responses, although at a relatively lower level. No differences in EAG responses to catnip oil were found between the two sexes of stable flies and three concentrations tested (Figure 6B). More interestingly, in the feedlot field studies, stable flies were observed to avoid the catnip treated areas by flying away abruptly (at least 5cm from the treated area). Such a behavior may further support a spatial repellent nature of catnip oil against stable flies.

Table 1. Spatial repellency of catnip oil and its compositional compounds in a single cage olfactometer against stable flies, *Stomoxys calcitrans* [a]

Treatments	*Significances*
Control (hexane) 74 ± 1 vs. Catnip oil (100 μg) 26 ± 1	$P < 0.001$
Control (hexane) 83 ± 2 vs. ZE-Nepetalactone (100 μg) 17 ± 1.5	$P < 0.005$
Control (hexane) 73 ± 2 vs. EZ-Nepetalactone (100 μg) 27 ± 2	$P < 0.005$
Control (hexane) 52 ± 8 vs. Caryophyllene (100 μg) 48 ± 8.4	$P = 0.85$
Control (hexane) 25 ± 5 vs. 1-Octen-3-ol (100 μg) 75 ± 4	$P < 0.01$
Control (hexane) 62 ± 4 vs. 1-Octen-3-ol +Catnip oil (100 μg) 38 ± 3	$P < 0.05$
Control (hexane) 60 ± 11 vs. DEET (100 μg) 40 ± 11	$P = 0.16$

[a] Results were mean percentage of stable flies observed in treated or control ports (± S.E., n = 60-80). Significances were measured using Student's T-test.

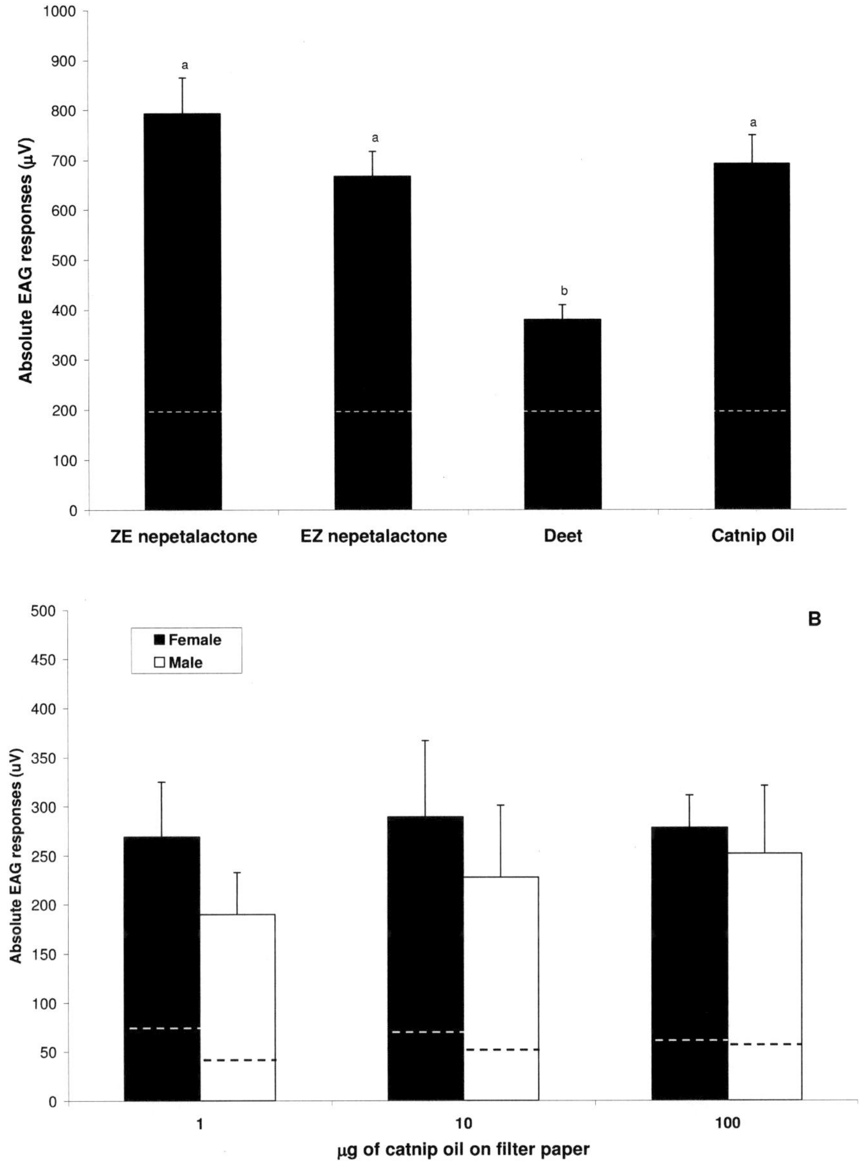

Figure 6. EAG recordings from stable flies responding to catnip oil and its compositional compounds (10 μg) (A, n = 12, 6 antennae of each sex), and stable flies responding to different dosages of catnip oil (B, antennae of male (n=8), female (n=12)). Dashed lines inside bars in figures were mean EAG responses to control puffs. No significant differences were found among the treatments, dosages and two sexes, P < 0.05, ANOVA.

Effectiveness of Catnip Oil Formulations against Stable Flies on Cattle

Although more than 95% repellency of catnip oil against stable flies was observed in the laboratory assays (*18*), its longevity was somewhat limited (*52*). Use of a 15% oil-based formulation topically applied onto cattle legs resulted in an effective repellency (>90%) that lasted up to 6 hours after the application (Figure 7A). The repellency disappeared 7 hours after the application. When a water-based catnip formulation (30%), was applied, the effective repellency (>90%) was only observed for 4-5 hours after the application (Figure 7B). Plant-based repellents with a high vapor pressure, such as catnip oil, may offer protection at low concentrations, but with the risk of loss of repellency in a short time. More work is necessary to discover new repellents with the same effectiveness in repellency for extended longevity. In addition, further research should be directed to explore more efficient formulation technologies via slow-release mechanisms. Although applications of catnip formulations for controlling stable flies in livestock industry may seem less practical due to cost, their use on pet animals against same flies should have some great potentials. In addition, the broad repellency of catnip oil against biting flies should further encourage the research to a significant level.

Oviposition Repellency of Catnip Oil

It has also been shown that catnip oil acts as an excellent ovipositional repellent against gravid stable flies. Results from laboratory oviposition assays demonstrate an inhibition rate of egg-laying over 97% (Figure 8A), when the oviposition media was mixed with 100 mg of catnip oil. A 4-choice oviposition experiment to compare repellency of catnip oil to its major constituent compounds (nepetalactones) has revealed that 500 times more eggs were laid on the untreated oviposition jars (< 100 eggs were found on controls), but no differences were found among the catnip/nepetalactones-treatments (Figure 8B). The strong oviposition repellency found from catnip oil may be used to develop sprayable formulations to apply in areas where female flies would be likely to oviposit, therefore, reducing their future populations.

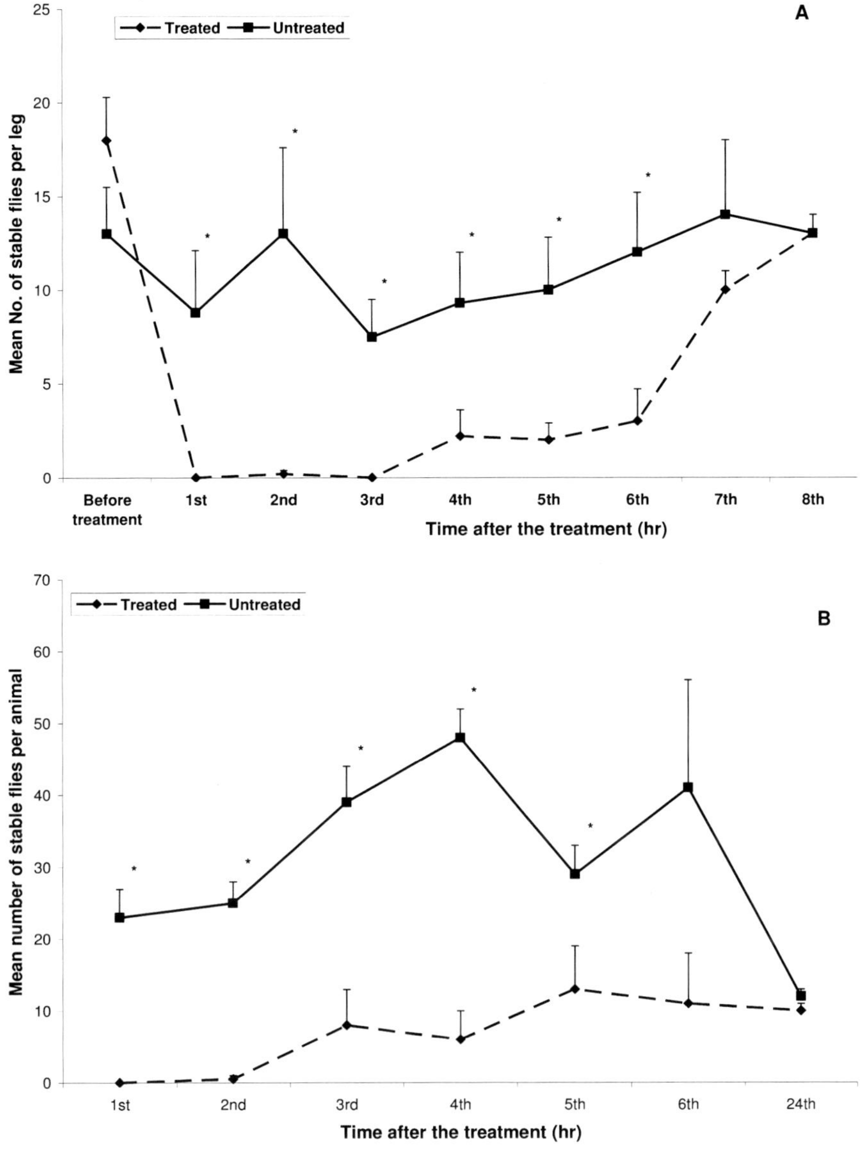

Figure 7. Mean number of adult stable flies observed landing on legs of cattle treated with (A) a 15% catnip oil-based formulation and the control; and (B) a 30% water-based formulation and the control. Means with an asterisk above squares of the solid line by time after treatment are significantly different at (P < 0.05, Student's T-test). Error bars show standard errors of the mean.

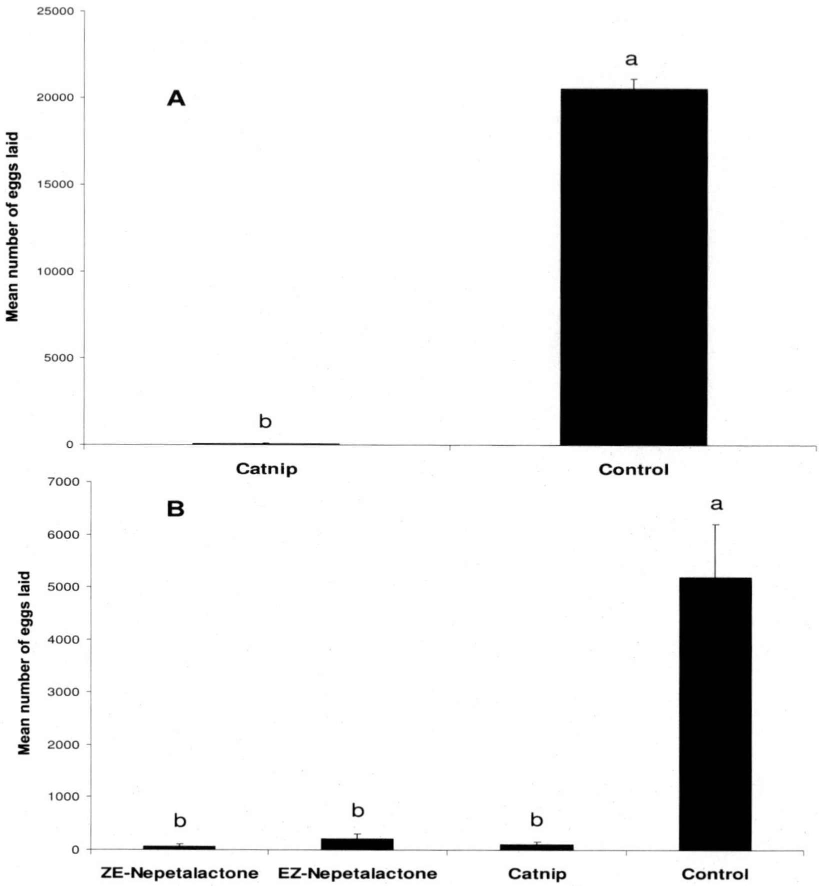

Figure 8. Mean number of eggs laid from oviposition jars treated with (A) 100 mg of catnip oil, or without; (B) 100 mg of catnip oil, two catnip components and the control. Means with different letters (A) above the bars are significantly different at P < 0.05 level (Student's T-test). Means with different letters (B) on top of the bars are significantly different at (P < 0.05, ANOVA followed by Duncan's test). Error bars show standard errors of the mean.

Repellent Activity of Catnip Oil against Other Filth Flies

Insect repellents have been used widely to protect humans and animals against arthropod attack. Among them, mosquito repellents have received the most attentions. For filth flies, so far only limited studies have been reported on repellent activities from plant materials (essential oils and short-chain fatty acids) and some synthetic insecticides (*18*, *52*, *59*, *60*). In addition to the repellency against the stable fly, catnip oil also repels other filth flies, such as house fly, horn fly and face fly, effectively (Figure 9).

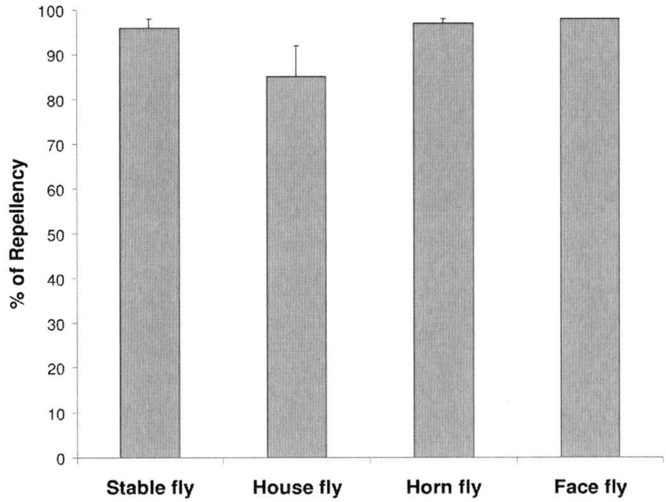

Figure 9. Mean percentages of repellency of catnip oil (20-mg concentration) observed from starved flies of 4 difference species.

Toxicity of Catnip Oil to Small Rodents

So far, toxicity tests have only been performed on two plant-associated repellent compounds, para-menthane-3,8-diol (derived from Australian lemon-scented gum tree) and picaridin (a synthetic derivative related to compounds in pepper) (US-EPA biopesticide registration documents 011550 and 7505C). Citronella oil, representing another group of botanical insect repellents is exempted from regulation under the Federal Insecticide, Fungicide, and Rodenticide Act of 1996 due to its very low or no toxicity (*15*). Based on the acute toxicity data of the three EPA-approved mosquito repellents, picaridin, para-menthane-3,8-diol and DEET, a comparative table containing toxicity data of catnip oil and above mentioned repellents has been summarized (Table 2). In general, acute toxicity of catnip oil appeared to be extremely low since almost no gross signs of toxicity were noted. Catnip oil may be the least toxic among the four repellents compared.

In summary, catnip oil and its major constituent compounds, nepetalactones, act not only as an efficient feeding and ovipositional repellent, but also have a strong spatial repellency. Field trials with two catnip oil formulations conducted on cattle gave at least 5-6 hours of protection against stable flies. Catnip oil formulated to meet USDA organic standards may also have promise as a method for stable fly control in organic dairy farms (*13*, *61*).

Table 2. Acute toxicity comparisons of several selected mosquito repellents

	DEET	Picaridin	p-Methane-3,8-diol	Catnip
Acute Oral	2170-3664 mg/kg	2236-4743 mg/kg	> 5000 mg/kg	2710-3690 mg/kg
Acute Dermal	>2000 mg/kg	>2000 mg/kg	>2000 mg/kg	>5000 mg/kg
Acute Inhalation	5.95 mg/L	>4364 mg/L	>2.17 mg/L	>10,000 mg/L
Primary Eye Irritation (72-hr)	Moderate	Moderate	Severe	Slight
Primary Skin Irritation	No	No	Slight	Slight
Dermal Sensitization	No	No	No	No

Acknowledgments

Drs. A. Zhang; C. Dunlap; X-P Zeng; D. Berkebile; B. Wienhold; M. Chaudhury and two anonymous reviewers provided materials, information and comments on the manuscript.

References

1. Taylor, D. B.; Moon, R. D.; Mark, D. R. Economic impact of stable fly (Diptera: Muscidae) on dairy and beef cattle production. *J. Med. Entomol.* **2010**, in press.
2. Zumpt, F. *The Stomoxyine Biting Flies of the World*; Gustav Fisher Verlag: Stuttgart, Germany, 1973; pp 1−175.
3. D'Amico, F.; Gouteux, J. P.; Le Gall, F.; Cuisance, D. Are stable flies (Diptera: Stomoxyinae) vectors of *Trypanosoma vivax* in the Central African Republic? *Vet. Res.* **1996**, *27*, 161–170.
4. Buxton, B. A.; Hinkle, N. C.; Schultz, R. D. Role of insects in the transmission of bovine leukosis virus: Potential for transmission by stable flies, horn flies, and tabanids. *Am. J. Vet. Res.* **1985**, *46*, 123–126.
5. Gilles, J.; David, J. F.; Duvallet, G. Temperature effects on development and survival of two stable flies, *Stomoxys calcitrans* and *S. niger niger* (Diptera: Muscidae), in La Réunion island. *J. Med. Entomol.* **2005**, *42*, 260–265.
6. Bishop, F. C. The stable fly *Stomoxys calcitrans* as an important livestock pest. *J. Econ. Entomol.* **1913**, *6*, 112–126.
7. Meyer, J. A.; Peterson, J. J. Characterization and seasonal distribution of breeding sites of stable flies and house flies (dipteral: Muscidae) on eastern Nebraska feedlots and dairies. *J. Econ. Entomol.* **1983**, *76*, 103–108.

8. Skoda, S. R.; Thomas, G. D.; Campbell, J. B. Developmental sites and relative abundance of immature stages of the stable fly (Diptera: Muscidae) in beef cattle feedlot pens in eastern Nebraska. *J. Econ. Entomol.* **1991**, *84*, 191–197.
9. Cilek, J. E.; Greene, G. L. Stable fly (Diptera: Muscidae) insecticide resistance in Kansas cattle feedlot. *J. Econ. Entomol.* **1994**, *87*, 275–279.
10. Marcon, P. C. R. G.; Thomas, G. D.; Siegfried, B. D.; Campbell, J. B. Susceptibility of stable flies (Diptera: Muscidae) from southern Nebraska beef cattle feedlots to selected insecticides and comparison of 3 bioassay techniques. *J. Econ. Entomol.* **1997**, *90*, 293–298.
11. Thacker, J. R. M. *An Introduction to Arthropod Pest Control*; Cambridge University Press: Cambridge, U.K., 2002; pp 1−360.
12. Ware, G. W.; Whitacre, D. M. *The Pesticide Book*, 6th ed.; Meister Media Worldwide: Willoughby, Ohio, 2004.
13. Isman, M. B. Botanical insecticides, deterrents, and repellents in modern agriculture and an increasingly regulated world. *Ann. Rev. Entomol.* **2006**, *51*, 45–66.
14. Barnard, D. R. Repellency of essential oils to mosquitoes (Diptera: Culicidae). *J. Med. Entomol.* **1999**, *36*, 625–629.
15. Sukumar, K.; Perich, M. J.; Boobar, L. R. Botanical derivatives in mosquito control: A review. *J. Am. Mosq. Control Assoc.* **1999**, *7*, 210–237.
16. Schultz, G.; Simbro, E.; Belden, J.; Zhu, J.; Coats, J. Catnip, *Nepeta cataria* (Lamiales: Lamiaceae) − A closer look: Seasonal occurrence of nepetalactone isomers and comparative repellency of three terpenoids to insects. *Environ. Entomol.* **2004**, *33*, 1562–1569.
17. Zhu, J.; Zeng, X.; Ma, Y.; Liu, T.; Qian, K.; Han, Y.; Xue, S.; Tucker, B.; Schultz, G.; Coats, J.; Rowley, W.; Zhang, A. Comparisons of adult repellency and larvicidal activity of plant essential oils against mosquitoes. *J. Am. Mosq. Control Assoc.* **2006**, *22*, 515–22.
18. Zhu, J.; Zeng, X.; Berkebile, D.; Du, H-J.; Tong, Y.; Qian, K. Efficacy and safety of a novel filth fly repellent. *Med. Vet. Entomol.* **2009**, *23*, 209–216.
19. Cook, A. M.; Khan, Z. R.; Pickett, J. A. The use of push−pull strategies in integrated pest management. *Annu. Rev. Entomol.* **2007**, *52*, 375–400.
20. Weseloh, R. M. Sense organs of the hyperparasite *Cheiloneurus noxius* (Hymenoptera: Encyrtidae) important in host selection processes. *Ann. Entomol. Soc. Am.* **1972**, *65*, 41–46.
21. Vinson, S. B.; Bin, F.; Strand, M. R. The role of the antennae and host factors in host selection behavior of *Trissolcus basalis* (Wall.) (Hymenoptera: Scelionidae). *Les Colloques de-I'INRA* **1986**, *43*, 267–273.
22. Bin, F.; Colazza, S.; Isidoro, N.; Solinas, M.; Vinson, S. B. Antennal chemosensilla and glands, and their possible meaning in the reproductive behavior of *Trissolcus basalis* (Woll) (Hymenoptera: Scelionidae). *Entomologica* **1989**, *30*, 33–97.
23. Castrejon-Gomez Victor, R.; RoJas Julio, C. Antennal sensilla of *Anastrepha serpentine* (Diptera: Tephritidae). *Ann. Entomol. Soc. Am.* **2009**, *102*, 310–316.

24. Sutcliffe, J. F.; Kokko, E. G.; Shipp, J. L. Transmission electron microscopic study of antennal sensilla of the female black fly, *Simulium arcticum* (Diptera: Simuliidae). *Can. J. Zool.* **1990**, *68*, 1443–1453.
25. Pfeil, R. M.; Walsh, R. A.; Mumma, R. O. Scanning electron microscopic examination of the putative olfactory structures possessed by the phorid fly, *Megaselia halterata* (Diptera: Phoridae). *Scanning Microsc.* **1994**, *8*, 687–694.
26. Shanbhag, S. R.; Singh, K.; Singh, R. N. Fine structure and primary sensor rojections of sensilla located in the sacculus of the antenna of *Drosophila melanogaster. Cell Tissue Res.* **1995**, *282*, 237–249.
27. Fernandes, F. F.; Pimenta, P. F. P.; Linardi, P. M. Antennal sensilla of the new world screwworm fly, *Cochliomyia hominivorax* (Diptera: Calliphoridae). *J. Med. Entomol.* **2004**, *41*, 545–551.
28. Sukontason, K.; Sukontason, K. L.; Piangjai, S.; Boonchu, N.; Chaiwong, T.; Ngern-klun, R.; Sripakdee, D.; Vogtsberger, R. C.; Olson, J. K. Antennal sensilla of some forensically important flies in families Calliphoridae, Sarcophagidae and Muscidae. *Micron* **2004**, *35*, 671–679.
29. Lewis, C. T. Structure and function in some external receptors. *Symp. R. Entomol. Soc. London* **1970**, *5*, 59–76.
30. White, S. L.; Bay, D. E. Olfactory sensilla of the Horn fly, *Haemaobia irritans* (L.) (Diptera: Muscidae). *J. Kansas Entomol. Soc.* **1980**, *53*, 641–652.
31. Bay, D. E.; Pitts, C. W. Antennal olfactory sensilla of the face fly, *Musca autumnalis* Degree (Diptera: Muscidae). *Int. J. Insect Morphol. Embryol.* **1976**, *5*, 1–16.
32. Krijgsman, B. J. Reizphysiologische Untersuchungen an blutsaugenden Arthropoden in Zusammenhang mit ihrer Nahrungswahl.-i. *Stomoxys calcitrans. ZeiUchr. f. vergl. Physiol.* **1930**, *11*, 702–29.
33. Hopkins, B. A. The probing response of *Stomoxys calcitrans* (L.), the stable fly, to vapors. *Anim. Behav.* **1964**, *12*, 513–524.
34. Zdarek, J.; Pospisil, J. Orientation of *Stomoxys calcitrans* L. towards warmth, during ontogenesis in relation to various food conditions. *Acta Entomol. Bohemoslov.* **1965**, *62*, 421–427.
35. Gatehouse, A. G. The Influence of Some Chemical and Physical Stimuli on the Feeding and Oviposition Behavior of *Stomoxys calcitrans* L. Ph.D. Thesis, University of London, 1969; pp 1−142.
36. Tangtrakulwanich, K.; Chen, H.; Baxendale, F.; Brewer, G.; Zhu, J. J. Characterization of olfactory sensilla of *Stomoxys calcitrans* and electrophysiological responses to odorant compounds associated with their host and oviposition media. *Med. Vet. Entomol.* **2010**, in press.
37. Lewis, C. T. Superficial sense organs of the antennae of the fly, *Stomoxys calcitrans. J. Insect Physiol.* **1971**, *17*, 449–461.
38. Logan, J. G.; Birkett, M. A. Semiochemicals for biting fly control: Their identification and exploitation. *Pest Manage. Sci.* **2007**, *63*, 647–657.
39. Jeanbourquin, P.; Guerin, P. M. Sensory and behavioral responses of the stable fly *Stomoxys calcitrans* to rumen volatiles. *Med. Vet. Entomol.* **2007**, *21*, 217–224.

40. Romero, A.; Broce, A.; Zurek, L. Role of bacteria in the oviposition behavior and larval development of stable flies. *Med. Vet. Entomol.* **2006**, *20*, 115–121.
41. Stephens, J. W. W.; Newstand, R. The anatomy of the proboscis of biting flies. Part II. *Stomoxys* (stable flies). *Ann. Trop. Med. Parasitol.* **1907**, *1*, 171–198.
42. Elzinga, R. J.; Broce, A. B. Labellar modifications of Muscomorpha flies (Diptera). *J. Kansas Entomol. Soc.* **1986**, *79*, 150–209.
43. Coats, J. R. Risks from natural versus synthetic insecticides. *Annu. Rev. Entomol.* **1994**, *39*, 489–515.
44. Mafong, E. A.; Kaplan, H. Insect repellents: What really works? *Postgrad. Med.* **1997**, *102* (2), 63–9.
45. Krajick, K. Keeping the bugs at bay. *Science* **2006**, *313*, 36–38.
46. Sharma, V. P.; Ansari, M. A.; Razdan, R. K. Mosquito repellent action of neem (*Azadirachta indica*) oil. *J. Am. Mosq. Control Assoc.* **1993**, *9*, 359–364.
47. Kant, R.; Bhatt, R. M. Field evaluation of mosquito repellent action of neem oil. *Ind. J. Malariol.* **1994**, *31*, 122.
48. Trongtokit, Y.; Rongsriyam, Y.; Komalamisra, N.; Apiwathnsorn, C. Comparative repellency of 38 essential oils against mosquito bites. *Phytother. Res.* **2005**, *19*, 303–309.
49. Peterson, C. J. Insect Repellents of Natural Origin: Catnip and Osage Orange. Ph.D. Dissertation, Iowa State University, 2001; pp 103−110.
50. Bernier, U. R.; Furman, K. D.; Kline, D. L.; Allan, S. A.; Barnard, D. Comparison of contact and spatial repellency of catnip oil and N,N-diethyl-3-methylbenzamide (DEET) against mosquitoes. *J. Med. Entomol.* **2005**, *42*, 306–311.
51. Eisner, T. Catnip: Its raison d'être. *Science* **1964**, *146*, 1318.
52. Zhu, J. J.; Dunlap, C.; Behle, R.; Berkebille, D.; Wienhold, B. Repellency of a wax-based catnip-oil formulation against stable flies. *J. Agric. Food Chem.* **2010**, *58* (23), 12320–12326.
53. Klun, J. A.; Debboun, M. A. New module for quantitative evaluation of repellent efficacy using human subjects. *J. Med. Entomol.* **2000**, *37*, 177–181.
54. Zhang, A.; Klun, J. A.; Wang, S.; Carroll, J. F.; Debboun, M. Isolongifolenone: A novel sesquiterpene repellent of ticks. *J. Med. Entomol.* **2009**, *46*, 100–106.
55. Klun, J. A.; Khrimian, A.; Margaryan, A.; Kramer, M.; Debboun, M. Synthesis and repellent efficacy of a new chiral piperidine analog: Comparison with DEET and Bayrepel activity in human-volunteer laboratory assays against *Aedes aegypti* and *Anopheles stephensi*. *J. Med. Entomol.* **2003**, *40*, 293–299.
56. Klun, J. A.; Strickman, D.; Rowton, E.; Williams, J.; Kramer, M.; Roberts, D.; Debboun, M. Comparative resistance of *Anopheles albimanus* and *Aedes aegypti* to N,N-diethyl-3-methylbenzamide (DEET) and 2-methylpiperidinyl-3-cyclohexen-1-carboxamide (AI3-37220) in

laboratory human-volunteer repellent assays. *J. Med. Entomol.* **2004**, *41*, 418–422.

57. Mihok, S.; Kang'ethe, E. K.; Githaiga, K. K. Trials of traps and attractants for *Stomoxys spp.* (Diptera: Muscidae). *J. Med. Entomol.* **1995**, *32*, 283–289.
58. Holloway, M. T. P.; Phelps, R. J. The responses of *Stomoxys spp.* (Diptera: Muscidae) to traps and artificial host odors in the field. *Bull. Entomol. Res.* **1991**, *81*, 51–55.
59. Hieu, T. T.; Kim, S.-I.; Lee, S.-G.; Ahn, Y.-J. Repellency to *Stomoxys calcitrans* (Diptera: Muscidae) of plant essential oils alone or in combination with *Calophyllum inophyllum* nut oil. *J. Med. Entomol.* **2010**, *47*, 575–580.
60. Mullens, B. A.; Reifenrath, W. G.; Butler, S. M. Laboratory trials of fatty acids as repellents or antifeedants against houseflies, horn flies and stable flies (Diptera: Muscidae). *Pest Manage. Sci.* **2009**, *65*, 1360–1366.
61. Isman, M. B. Botanical insecticides: For richer, for poorer. *Pest Manage. Sci* **2008**, *64*, 8–11.

Chapter 7

Using Lone Star Ticks, *Amblyomma americanum* (Acari: Ixodidae), in *in Vitro* Laboratory Bioassays of Repellents: Dimensions, Duration, and Variability

J. F. Carroll,*,[1] A. Zhang,[1] and M. Kramer[2]

[1]Invasive Insect Biocontrol and Behavior Laboratory, Beltsville Agricultural Research Center, Agricultural Research Service, U.S. Department of Agriculture, Beltsville, Maryland 20705
[2]Biometrical Consulting Service, Beltsville Agricultural Research Center, Agricultural Research Service, U.S. Department of Agriculture, Beltsville, Maryland 20705
***E-mail: john.carroll@ars.usda.gov**

The *in vitro* laboratory bioassay is an important tool in tick repellent discovery and development, with a variety of bioassays used in recent years. Several factors, such as size and configuration of test surfaces and duration of tick exposure, can influence the outcome of bioassays. We tested two tick repellents, *N,N*-diethyl-3-methyl benzamide (deet) and (-)-isolongifolenone, in seven different bioassays or configurations. All bioassays used ≥4 concentrations of repellent and an ethanol control applied to filter paper against lone star tick nymphs, *Amblyomma americanum* (L.). Climbing bioassays included a 22 × 1 cm vertical filter paper strip and a 4 × 7 cm vertical filter paper strip plus four modifications of the basic 4 × 7 cm configuration. We used a moving object bioassay (MOB), in which a strip of filter paper treated with test solution was affixed to a rotating heated brass drum and ticks allowed to transfer to the paper. A horizontal bioassay in which ticks were confined between two filter paper discs that had one half treated with repellent was also used. For each bioassay, deet and (-)-isolongifolenone were similarly effective, but in some bioassays ticks were repelled by lower concentrations of both repellents than in other bioassays. The 22 × 1 cm strip proved impractical for regular bioassay use, but showed that a height of 8-9 cm and ~6 min duration were optimal for climbing bioassays. When a loop of treated paper was added to untreated lower portion of the 4 × 7 cm filter paper, as alternative escape for ticks responding to repellents, more ticks were on the loop and lower untreated area of the strip at 10 min (end of the test) than were on the lower untreated area of the basic 4 × 7 cm strip. However, with the ethanol controls more ticks fell from 4 × 7 cm strips with loops than those without loops. Several important behaviors associated with host acquisition (contacting, transferring to and remaining on a moving surface) were recorded in the MOB, but we only found significant differences between treatment and control for the proportion of ticks that transferred to the filter paper and the length of time the ticks remained on paper. The petri dish bioassays lasted longer than other bioassays (2h compared to 10 min for the vertical filter bioassays) and allowed detection of a decline in repellency over time. Individual variation among ticks and fatigue (change in response) in repeatedly tested ticks were assessed in a vertical paper strip bioassay using deet. The responses of ticks tested twice on one day (morning and afternoon) did not differ between tests. However, continued repeated daily testing compromised results. A hiatus of about a week between tests allowed ticks to return to their initial response profiles.

Keywords: deet; (-)-isolongifolenone; dose response; repellency

Tick-borne diseases are a serious and increasing problem in United States and elsewhere in the habitable world (*1*). A variety of tick control measures have been developed and implemented (*2*), but repellents remain an important means of personal protection against tick bite (*3*). Repellent products, such as deet and permethrin, used on skin and clothes respectively, have been available for decades. However, there is a rising demand for novel, effective, safe, inexpensive tick repellents (*4*). The recent discovery of olfactory receptor neurons for repellents in *Drosophila* may lead to novel approaches for repellent testing (*5*), but *in vitro* and/or *in vivo* behavioral bioassays will probably remain a fixture in the discovery, development and registration of repellents for the foreseeable future. Behavioral bioassays should yield reliable, meaningful data that accurately represent the efficacy of a test compound or essential oil.

The lone star tick, *Amblyomma americanum* (L.), has grown in importance as a nuisance biter and vector of pathogens, such as *Ehrlichia chaffeensis* Anderson, Dawson, Jones and Wilson, the causative agent of human monocytic ehrlichiosis (*6*). Stromdahl et al. (*7*) reported a high prevalence of spotted fever group rickettsiae in lone star ticks from Maryland. *Amblyomma americanum* occurs from the south-central and southeastern United States northward along the Atlantic seaboard to New England (*8*). The distribution of *A. americanum* has been expanding northward along the Atlantic Coast (*6*, *9*, *10*). Although *A. americanum* lacks the cachet and attention of the blacklegged tick, *Ixodes scapularis* Say, the principal vector of the Lyme disease pathogen, there are some advantages to using *A. americanum* in repellent bioassays. First, it is easier to rear *A. americanum* on a large scale than *I. scapularis*, so the former are obtainable in greater quantities and, if purchased, at lower prices. Second, behavioral bioassays of repellents depend on the arthropod subjects moving about; A. americanum do so more readily and rapidly than I. scapularis. Although *A. americanum* are active host seekers whose strategy tends toward the hunter type (*11*), in nature host contact may often occur while ticks are on questing sites on vegetation. Unlike *I. scapularis*, however, *A. americanum* readily abandon questing sites and will move several meters toward a host. The lone star tick is well known for its proclivity to move rapidly toward sources of CO_2 (*12*). Laboratory-reared *A. americanum* nymphs appear to be suitable replacements for field-collected nymphs, as demonstrated by Carroll et al. (*13*) who found that laboratory-reared nymphs from Texas and Oklahoma responded similarly to field collected nymphs from Maryland in dose response bioassays using deet and racemic 220.

The characteristic responses of *A. americanum* to repellents are epitomized in the bioassays reported by Carroll et al. (*14*), in which *A. americanum* and *I. scapularis* nymphs were subjected to the same tests using deet and SS220. When host-seeking *A. americanum* nymphs were encircled by a 1-cm wide ring of test solution on a horizontal filter paper disc, they routinely crossed concentrations of deet and SS220 that repelled all *I. scapularis* nymphs, confining the latter within the repellent-treated ring. However, concentrations that did not repel *A. americanum* nymphs on the horizontal filter paper, repelled them on a vertical surface from which they could drop. When the middle 4 × 5 cm of a 4 × 7-cm filter paper strip was treated with test solution, and the paper dried and suspended vertically, ticks were allowed to mount the lower untreated edge. As depicted in Carroll et al. (*14*), the dose response curve of *A. americanum* nymphs to deet in the vertical bioassay slopes gradually compared to the steep curve for *I. scapularis* to deet. Many *A. americanum* dropped from the vertical papers treated with repellent, whereas *I. scapularis* would either not enter the treated portion of the vertical paper or shortly after entering retreat to the lower untreated zone.

This difference in the behavior of *A. americanum* and *I. scapularis* was also observed in fingertip tests with elemol (*15*). In responding to repellents, few *A. americanum* tend to remain near but not on the treated surface (*15*). Instead, they crawl away or release their hold on a vertical surface and fall. When an *A. americanum* nymph rushes onto a barrier treatment a few centimeters wide, there is some chance that it might continue completely across the treatment because it can no longer detect a repellent gradient associated with the edge of the treatment.

However, a tick that tends to approach the repellent slowly and penetrates the treatment only slightly if at all, is less likely to cross a barrier treatment by chance. Physical and temporal parameters, such as the width of barrier treatments (the distance a tick must cross to be considered not repelled), influence the outcome of tick repellent bioassays. Sometimes ticks may cross a repellent-treated surface only after entering and retreating a few times, so a bioassay that ends without allowing a sufficient yet reasonable time for a tick to reencounter the repellent would overestimate the repellent's protective capacity.

Variation, perhaps associated with the "dash through or drop" reaction of *A. americanum* to repellents, is observed less often in bioassays when weakly or strongly repellent test solutions are tested, but is manifested in dose response studies. For example, dose response results for (-)-isolongifolenone and deet in fingertip bioassays against *A. americanum* nymphs were similar (*16*), but the results for (-)-isolongifolenone were notably more variable, with some higher concentrations repelling fewer ticks than lower ones. The sesquiterpene (-)-isolongifolenone occurs naturally in *Humiria balsamifera* St. (Aubl.) Hill (Humiriaceae), a tree found in South America (*17*) and is dissimilar in structure from deet. Variation in responses is expected, but excessive variation requires extra replicates and muddles interpretation of bioassay outcomes. Excess variability can limit which different compound/concentrations can be discriminated.

Acknowledging variation in behavior among tick species, to chose to keep keep matters simple and compare results from different bioassay systems using the same tick species and life stage. We examined the responses of *A. americanum* to two repellents, deet and (-)-isolongifolenone, in several bioassays to ascertain the strengths, weaknesses, and reliability of the various methods and to define optima for test time and physical dimensions. Specifically, we wanted to answer the following questions: (1) in vertical filter paper tests, how long should the paper strip be?; how long should the test last?, (2) does adding a bottom loop to vertical filter paper tests help prevent ticks from dropping off?; if so, is it better for the ticks to climb onto the strip (loop) near the part with the repellent challenge or further from it?, (3) how do moving object bioassays compare to other repellent bioassays for ticks?, (4) how do choice experiments (ticks confined to a petri dish where they must choose between substrates with and without a repellent) compare with other repellent tick bioassays?, and (5) since purchasing ticks is expensive, can they be reused in repellent bioassays?

Methods

Ticks

Host-seeking *A. americanum* nymphs were obtained from colonies at the USDA, ARS, Knipling-Bushland U. S. Livestock Insects Research Laboratory, Kerrville, TX and Oklahoma State University, Stillwater, OK. The ticks were held at 23-24° C, ~97% RH and a photoperiod of 16:8 h (L:D), and tested 3-6 mo after molting.

Chemicals

(-)-Isolongifolenone was efficiently prepared as a sole major product from (-)-isolongifolene (Sigma, St. Louis, MO) utilizing *tert*-butyl hydroperoxide as the oxidants, chromium hexacarbonyl as the catalyst, and acetonitrile and benzene as the solvent in high isolated yield (≥90%) with high purity (≥99%) in a short reaction time (~2h) (Wang and Zhang 2008). Deet was purchased from Aldrich, Sigma-Aldrich, St. Louis, MO. 95% Ethanol (Sigma-Aldrich, St. Louis, MO) was used as the blank control and the solvent to make deet and (-)-isolongifolenone solutions for the assays.

Composite Scores

A method we developed (*18*) to optimally combine the various behaviors typically exhibited by ticks as they navigate a test paper strip into a single score was used on the moving object bioassay experiment described below, and works well when many concurrent (behavioral) measures are taken on each individual animal in an experiment and one wants to create a single composite score for the individual animal. An outline of this method follows, detailed information is provided in Kramer et al. (*18*). The basic idea is to use the behavioral differences observed as ticks are tested on different compounds to find optimal weightings of these behaviors (that best discriminate among the compounds) using canonical discriminant analysis. Compounds to which ticks responded similarly (in theory, compounds that ticks do not discriminate between) will produce similar composite scores, those where behaviors differed will have different scores.

In addition to variables measuring duration or counts of behaviors, indicator variables were created with a value of 1, if the behavior was performed, and 0, if not. This was done so that all variables could be included in the analysis, even if not performed by all ticks. Useful variables to create the scores were determined in a stepwise discriminant selection procedure. One dimensional composite scores were created by first fitting canonical discriminant functions, which consisted of the sum of these variables with weights (referred to as 'loadings') that best separated the compounds, and using scores from the first canonical discriminant function. Although, in theory, the scores could have more than one dimension (or axis), in no case did we find more than the first discriminant axis was useful. Thus, a composite score was created for each individual tick, and it consisted of a single number.

Experiments on the repeated testing of ticks also made use of composite scores, though the loadings used came from an earlier study (see 'Variation and repeated use of ticks' below). Part of this methodology was used in the '22-cm filter paper strip' experiment (see below) to identify behaviors that discriminated among the compound-concentration combinations, although the final creation of the composite score was not necessary.

22 × 1-cm Vertical Filter Paper

A 1 × 22-cm strip of Whatman No. 4 filter paper was marked with a lead pencil at 1 cm intervals and 165 µl of test solution evenly applied by pipettor to all but the terminal 1-cm sections. Concentrations of 103, 206, 413, 825, 1238, 1650 nmol deet or (-)-isolongifolenone/cm^2 and an ethanol control were tested. The strip was allowed to dry for ~10 min, and suspended vertically from a bulldog clip attached to a clip on a work holder (Aptex Corp., Bethel, Connecticut). A vial containing ticks was opened in a moated petri dish. An active (crawling or waving its forelegs) tick was allowed to mount the lower untreated end of the strip by holding the vial close to the filter paper or letting a tick mount a section of bamboo barbeque skewer from which the tick transferred to the filter paper. The locations of the tick were recorded at 1-min intervals, as were whether it dropped from the strip or climbed through the area that received the repellent or ethanol treatment. The time and location at which a tick fell from the strip and the highest location the tick attained were also recorded. A strip was reused with other ticks until 30 min after the first tick climbed on the strip. A moated petri dish beneath the strip confined ticks that fell from the strip. Twenty nymphs were tested for each concentration of (-)-isolongifolenone and deet. Nymphs were tested with an ethanol control each day repellents were tested.

The interest in this experiment was to determine both the optimal time for a single trial and the optimal length of the paper strip. We employed a stepwise discriminant analysis (SAS Proc Stepdisc) for both variables using modified data sets (similar to the methodology used to create composite scores). These were created using a Perl program which recoded the data as if the paper strip had been shorter. For example, if the paper strip sheet had been 10 cm rather than 20 cm and a tick dropped off at the 12 cm mark, that tick would have been recorded as completing the test by walking to the top of the strip. For each potential height (starting at 3 cm to the full 20 cm) we noted the average squared canonical correlation (these increase with improved discrimination) with the set of variables selected by the stepwise procedure (these sets could be different for different heights) and which location times (tick location at 1 min, 2 min., etc.) were most often included. We also analyzed subsets of the data (by dropping one of the compound-concentration combinations in turn) to make sure that results were not driven by a single combination.

4 × 7-cm Vertical Filter Paper

A 4 × 7-cm strip of Whatman No. 4 filter paper was marked with a line 1 cm from and parallel to each end. The area between the lines (4 × 5 cm) received 165 µl of test solution evenly distributed by pipettor and was allowed to dry for ~10 min. Concentrations of 206, 413, 825, and 1650 nmol deet or (-)-isolongifolenone/cm^2 filter paper and an ethanol control were tested. The strip was suspended vertically from a bulldog clip attached to a clip on a work holder over a moated petri dish. A vial containing ticks was opened in a moated petri dish and 10 nymphs were allowed to climb onto the lower untreated edge of the filter paper. As the situation dictated, the vial was held close to the filter paper or the filter paper (attached to the bulldog clip) was held close the vial in the petri dish to allow ticks to transfer. Tick locations were recorded at 1, 3, 5, 10 and 15 min after the tenth tick mounted the filter paper. Ticks were considered repelled if they fell from the filter paper without having crossed into the upper untreated area or were on the lower untreated area at 15 min after the tenth tick mounted the filter paper. Three replicates of 10 nymphs each were tested for each concentration of (-)-isolongifolenone and deet and ethanol control.

We tested the proportion of ticks repelled at the end of each trial for differences between the two compounds by fitting a generalized linear model, assuming that the proportions were samples from an over-dispersed binomial distribution, where the dependent variable is modeled as the logit of the proportion repelled. We used a square root transformation on concentration as it produced a more linear relationship with the logit of proportion repelled. We tested for difference between the two compounds by including a compound and compound by concentration interaction terms in the model, and noted the estimated over-dispersion for the compounds modeled separately.

4 × 7-cm Filter Paper with Loop

In order to provide *A. americanum* nymphs a third option (other options are dropping off or remaining in untreated area) for responding to a repellent barrier, the basic 4 × 7-cm filter paper was modified with two lateral extensions (1 × 6 cm, 1 × 5 cm) of the lower untreated zone that were curved to overlap 1 cm and were joined with transparent tape forming a ring or loop (Figure 1). The loop allowed ticks to move away from the 4 × 5-cm treated area with the possibility of returning and repeatedly challenging the repellent barrier. Two configurations of the avoidance loop were used. In both configurations, the upper 1 × 4 cm of the rectangle and the loop were untreated, as was a 1 × 4 cm approach tab that extended below the level of the loop. Ticks were allowed to climb onto the approach tab to start the bioassay. In the first configuration (Figure 1, panel A), the approach tab was directly below the 4 × 5 cm treated area and ticks could climb in a completely vertical route without interruption. In the second configuration (Figure 1, panel B) the tab was offset so that the left lateral margin of the approach tab was almost directly below the right margin of the rectangle. With the offset tab, ticks had to adjust their paths to continue to ascend, perhaps slowing their momentum as they encountered the treated section of the filter paper. The second configuration was tested with the loop on the near side and on the opposite side of the rectangle to the investigator, who was seated 0.6 m distant. Tick locations were recorded as in the 4 × 7-cm filter paper test. Three replicates of 10 nymphs each were tested for each concentration (206, 413, 825, and 1650 nmol/cm^2 filter paper) of (-)-isolongifolenone or deet, and an ethanol control.

We followed methodology similar to that given for the 4 × 7-cm filter paper trials to test for compound differences for each of the three paper configurations. In addition, we merged the datasets and, after removing the control (ethanol) trials, developed a generalized model to fit the proportion of ticks that fell (main effects were paper configuration, concentration, and compound). We used the step function in R for this (similar to stepwise regression, using an AIC estimate to determine relative model fit). We also looked to see if there were differences in the control proportions that fell for the different paper configurations.

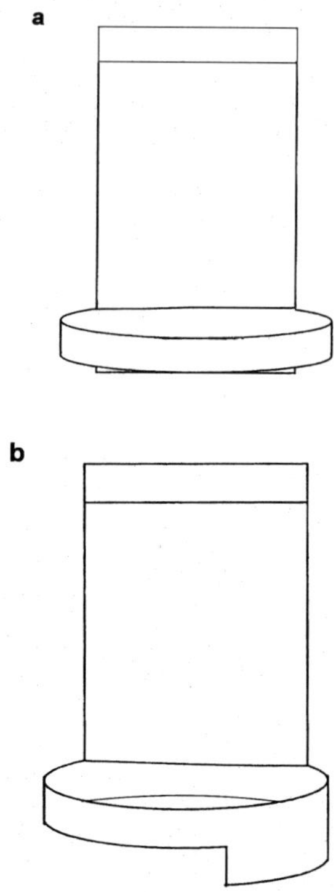

Figure 1. Loop configurations based on 4 × 7-cm vertical filter paper. A) Loop (1 cm wide, ~4 cm diam) with 1 cm extension directly below 4 × 7-cm rectangle. B) Loop same dimensions as A, but with extension offset so that ascending ticks could not go directly onto rectangle. Test solutions were applied to area between horizontal lines 1 cm from top and bottom of rectangle.

Moving Object Bioassay (MOB)

The moving object bioassay (MOB), described in detail by Dautel et al. (*19*, *20*), is an *in vitro* system that features heat and motion, stimuli associated with the presence of a host. The system has been used primarily against *Ixodes* spp. (*21*). Douglas et al. (*22*) used a version of it with *A. americanum* nymphs. Briefly, a brass cylinder (drum) contained water warmed by an immersion heater (Tempco, Wood Dale, IL) heated to maintain temperatures of ~34-36°C on the drum's outer surface. A rotisserie motor rotated the drum horizontally at 13 -15 rpm. A strip of Whatman No. 4 filter paper with a 2 × 10 cm section treated with 165 µl was affixed closely to the side of the cylinder over a brass plate soldered in place. Concentrations of 206, 413, 825, and 1650 nmol deet or (-)-isolongifolenone/cm^2

filter paper, and an ethanol control were tested. The plate caused the paper to protrude slightly from the surface of the drum. A petri dish half (2 cm deep, 6 cm diam) containing a silicone island and water (to confine ticks to the island) was held in place at the level of the drum. An inverted L-shaped wire projected from the silicone island. A small platform fashioned from clay for placement of a tick was affixed at the bend of the wire. The tip of the wire (nearly perpendicular to the side of the drum) was positioned 1-2 mm from the surface of the filter paper, just close enough for a nymph to catch hold of the filter paper with its forelegs as the paper passed by on the rotating drum. We recorded whether a tick contacted the filter paper, transferred to the paper and dropped from the paper. The time elapsed until the tick reached a mark 1 cm from the tip of the wire, reached the tip of the wire, transferred to the filter paper and crawled or dropped off the paper were also recorded.

We used the composite score method, explained above, to create linear discriminate functions that best separated the compound-concentration combinations. Because we found very poor separation, we tried a number of modifications by subsetting the data to improve the separation. Since the resulting composite scores appeared to be close to normally distributed, we used ANOVA to estimate which compounds differed and to estimate R^2.

Petri Dish Choice Bioassay

One half of each of two Whatman No. 4 filter paper discs (9.0 cm diam) marked into halves with a lead pencil was evenly treated (by pipettor) with 200 µl ethanol, which was allowed to dry for 10-15 min. When the ethanol application dried, an equal volume of test solution was applied to the other half of each filter paper disc and was allowed to dry for 10-15 min. Concentrations of 157, 315, 629, and 1258 nmol deet or (-)-isolongifolenone/cm^2 filter paper and an ethanol control were tested. One filter paper disc was placed in a disposable plastic petri dish lid (9.3 cm diam). A piece of wire (1.0 cm long, 0.1 cm diam) was placed on the ethanol treated half of the filter paper disc and similar piece of wire on the repellent-treated half of the disc. Five nymphs were dumped from a Fluon™-coated centrifuge tube (0.4 cm inner diam, truncated to a length of 3.5 cm) on a disc of parafilm (0.6 cm diam) affixed by pressure on the center point of the filter paper. A second filter paper disc was placed on top of the disc in the petri dish, so that repellent and ethanol treated halves aligned. A Mason jar (0.94 l) lid ring (8.8 cm outer diam) placed on the filter paper and held in place by two rubber bands confined the ticks between the filter papers, a method used by Crystal and Demilo (*23*) to confine mites in toxicant bioassays. The locations of the ticks were recorded at 10, 30, 60 and 120 min after the ticks were released on the filter papers. To aid in counting the ticks between the filter paper discs and discerning the diameter line, a flashlight (0.15 m distant) beam was shone briefly through the layers of paper.

For analysis, we used methodology similar to that described above, fitting a generalized linear model based on a quasi-binomial (over-dispersed binomial) distribution, and estimating means and a 95% confidence interval about the mean for each compound-concentration combination at each of the four time points.

Variation and Repeated Use of Ticks

To assess variation among ticks in how they respond to repellents, individually identified ticks were tested repeatedly within a day and over days in a vertical paper strip bioassay similar to those described above. Unlike the other bioassays we describe which used filter paper, this bioassay used recycled bond paper. Briefly, 15μl of test solution was applied evenly with a pipettor to the area (4 cm^2) between the 2 and 6-cm marks of a 1 × 8-cm strip of paper marked transversely at 1-cm intervals. Acetone was the solvent. The concentration (0.016 mg deet/cm^2 paper) was determined by preliminary testing to repel 40-60% of the ticks. After the paper had dried for 10 min, it was suspended vertically, and a tick was allowed to crawl onto the lower untreated portion of the strip. Observations lasted until the tick climbed past 6 cm, fell from the paper without climbing past 6 cm or 10 min elapsed from the time the tick crawled onto the paper. The behaviors recorded (some are presence/absence, some are duration) are listed in Table 1. One group (n = 15) of *A. americanum* nymphs was tested twice a day for three consecutive days. A second group (n = 15) of nymphs was tested twice a day for four consecutive days and, after a hiatus of 3 d, tested twice a day for two consecutive days. Thirty nymphs were tested once a day and at intervals of 5, 13, 3, and 4 d thereafter.

Table 1. *In vitro* bioassays discussed in this chapter. All had test solutions applied to Whatman No. 4 filter paper

22 x 1-cm vertical filter paper strip
4 × 7-cm vertical filter paper strip
4 × 7-cm vertical filter paper strip, loop extended, direct
4 × 7-cm vertical filter paper strip, loop extended, offset near observer
4 × 7-cm vertical filter paper strip, loop extended, offset near observer
moving object bioassay (MOB)
petri dish choice

Although several behaviors were recorded during a trial, since ticks were always tested with the same concentration of repellent, we could not employ the methods in Kramer et al. (*18*) which used different compounds to create a composite score. Instead, we created the composite score from the weights (loadings) used in Weldon et al. ((*24*), Table 1 in that paper), which used the same testing method and produced clear discrimination among many compounds, ranging from those with little repellent activity to those with considerable repellent activity. We reasoned that if a tick's performance deteriorated over time by repeated testing, it would show similar changes to being tested on a more effective repellent. For example, after many tests it might be more likely to drop off the paper strip earlier, be more reluctant to cross the area with repellent, etc. Preliminary analyses suggested that, at least for ticks tested frequently over many days, the composite score was a good summary, with values increasing with the number of repeated tests (consistent with increasing values for more repellent compounds in Weldon et al. (*24*).

We fit the data with mixed models using the nlme package in R (*25*), with the composite score as the dependent variable, and test day, time of day (AM or PM) as fixed independent variables, and individual tick as a random block effect. Test day was treated as either a regression variable (with a linear and quadratic component) or as a factor; typically a better fit (judged using AIC) resulted from using test day as a regressor. The basic model was altered as appropriate for the different experiments (e.g. in one of the experiments ticks were tested only in the morning). Residuals were inspected for autocorrelation (for an individual tick, it is possible that residuals from sequential trials would be more alike than residuals separated by more time), but none was found.

Results

For each type of bioassay, deet and (-)-isolongifolenone were similarly repellent to *A. americanum*, indicating that for the purpose of comparing the efficacy of the two compounds (deet generally considered the standard of repellent activity), the various filter paper bioassays yielded the same conclusions.

22 × 1-cm Vertical Filter Paper

We found that the optimal height of the filter paper strip for testing *A. americanum* was approximately 8-9 cm, which resulted in the highest canonical correlation (Table 2), with correlations decreasing as one moved away from that distance. This was the optimal paper strip height for all subsets of compounds, as well as the full set. We found that tick locations after 6 min were not selected for the 8-9 cm height, and rarely selected for other heights. Tick locations at 6 min were marginally or not significant (though selected to be in the model), so perhaps tests could be even shorter.

4 × 7-cm Vertical Filter Paper

We found no statistical difference in the slope of deet and (-)-isolongifolenone for the logit of the proportion of repelled ticks regressed on the square root of concentration ($p = 0.423$, *t*-test, 26 d.f.). Thus, the two compounds appear to have similar repellent activity at the same concentrations (regression equation: logit $(p) = -3.799\ [0.562] + 0.106\ [0.018] \times \text{sqrt (conc.)}$, standard error of estimates in square brackets, concentration in nmol/cm^2, p is the proportion repelled) (Figure 2 illustrates the data and fitted model). The over-dispersion parameter was larger for deet (2.239 versus 1.287), though both are well within the range commonly seen for experiments of this kind, with responses to deet indicating moderate over-dispersion.

Table 2. Results from a stepwise selection on useful behaviors to discriminate among compounds for filter paper strip heights of 6-10 cm. Behaviors abbreviations are: Loc*x*min = tick location at *x* min, l*x*d = 0 if tick was still on strip at *x* min, 1 if tick dropped off the strip at or before *x* min, DropLoc = height (cm) where tick dropped off strip

Height (cm)	*Behaviors, in order of entry*	*Average squared canonical correlation*
6	Loc1min, Loc5min, l4d	0.0560
7	Loc1min, Loc1min, l1d, Loc5min	0.0644
8	Loc1min, Loc3min, Loc2min, Loc6min, l5d	0.0762
9	Loc1min, Loc5min, Loc3min, l2d, DropLoc	0.0797
10	Loc1min, Loc5min, DropLoc	0.0628

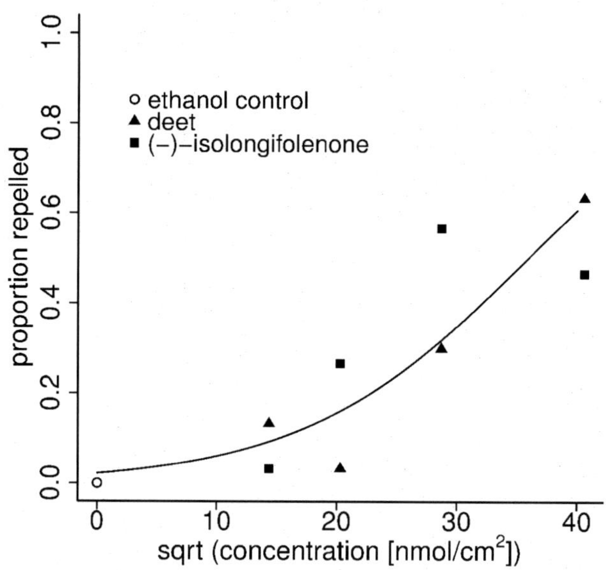

Figure 2. Points represent the proportion of ticks (aggregated over trials) that did not successfully crawl above the treated area in the 4 × 7-cm vertical paper test. The line represents the model fit to these data (note: this is a straight line on the logit scale).

4 × 7-cm Vertical Filter Paper with Loop

The data (proportion repelled and proportion that fell off the paper) are illustrated in Figure 3 for all three kinds of added loops and the two repellent compounds. Results for the paper configuration with no offset above the ring ('direct') were very similar to those of the previous experiment; no significant differences (p = 0.428, t-test, 28 d.f.) were found between the two compounds for the number of ticks repelled (regression equation: logit (p) = -1.631 [0.280] + 0.0837 [0.0122] × sqrt (conc.), standard error of estimates in square brackets, concentration in nmol/cm^2). In this experiment, the over-dispersion parameter was smaller for deet (1.040 versus 1.530). Results for the paper configuration where the offset was near the researcher were similar to those from no offset; no significant differences (p = 0.192, t-test, 35 d.f.) were found between the two compounds (regression equation: logit (p) = -0.337 [0.194] + 0.0541 [0.0103] × sqrt (conc.), standard error of estimates in square brackets, concentration in nmol/cm^2). In this experiment, the over-dispersion parameter was larger for deet (1.749 versus 1.174), Similar results were again obtained when the offset was opposite the researcher (far), (regression equation: logit (p) = -2.087 [0.440] + 0.0808 [0.0178] × sqrt (conc.), standard error of estimates in square brackets, concentration in nmol/cm^2). In this experiment, the over-dispersion parameter was larger for deet (3.816 versus 2.384),

The data sets were combined to determine if there were differences in the proportion of ticks that fell, and a higher dimension model was fit with a stepwise procedure. The model produced suggested that the offset paper configuration, with the researcher far from the loop, differed from the other two configurations in that far fewer ticks fell at lower concentrations (Figure 3, panel B), but with a more positive slope (so that falling rates were similar at high concentrations). There was also a systematic larger difference (about 3 times as large, on the logit scale) in falling rates between the direct and near configurations for (-)-isolongifolenone than for deet (i.e. on Figure 3, panel B, the points for (-)-isolongifolenone for the direct and near paper configurations are mostly far apart at the same concentrations). However, there was no significant difference between the paper configurations for the ethanol controls (p = 0.100, t-test, 13 d. f., over-dispersion parameter = 2.210).

For the two highest doses, 825 and 1650 nmol/cm^2 filter paper, of (-)-isolongifolenone, proportions of 0.17 and 0.07 ticks (n = 30) remained below the treatment at 10 min in the offset ring (near) compared to proportions of 0.03 and 0 ticks (n = 30) in the basic 4 × 7-cm bioassay. For the same doses of deet, proportions of 0.13 and 0.17 ticks (n = 30) remained below the treatment in the offset loop (near) compared to proportions of 0.07 and 0.10 ticks (n = 30) in the basic 4 × 7-cm bioassay. No ticks (n = 30) fell from ethanol controls of the basic 4 × 7-cm bioassay, whereas proportions of 0.15 (n = 40), 0.40 (n = 70), and 0.18 (n = 50) ticks fell from the controls of the direct, offset (near), and offset (far) ring 4 × 7-cm bioassays respectively.

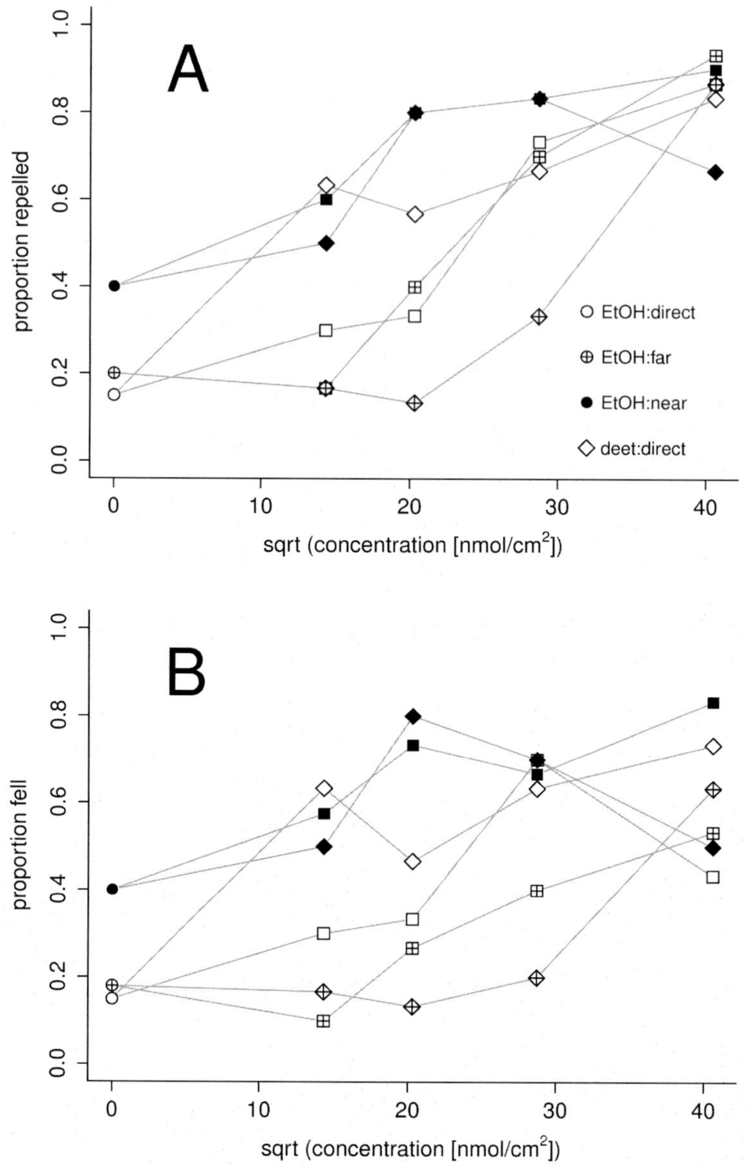

Figure 3. Points represent the proportion of ticks (aggregated over trials) that did not successfully crawl above the treated area (panel A) or fell (panel B) in 4 × 7-cm vertical paper tests with three different types of added loops and two compounds. The gray lines are an aid for following individual compound-loop combinations over the concentrations. Without including the effect of over-dispersion, the s.e. for each point would be 0.1. Thirty ticks were tested for each compound concentration and control.

Thus, there is an effect of the configuration of the paper used in these tests. All added loop configurations tended to increase drop rates of ticks, even in control conditions. This affects the proportion 'repelled', since a tick that drops is considered to be 'repelled', as one can readily observe by noting the similarity between the two panels in Figure 3. Of the three configurations with loops, the best appears to be 'far', but this configuration does not seem to improve on the vertical paper without a loop in vertical repellent tests.

Moving Object Bioassay (MOB)

Results from following the composite score methodology (Table 3) yielded composite scores that were close to normally distributed but also with means close together (i.e. there was not a linear discriminant function that, based on the behaviors observed, could separate the compound-concentration combinations). Only deet at 413 nmol/cm^2 was significantly different than the ethanol control, and the ranked means did not correspond to the concentrations (which makes little sense). We then redid the composite scores using fewer concentrations (e.g. ethanol, deet at 1650 nmol/cm^2 filter paper, (-)-isolongifolenone at 1650 nmol/cm^2), then applying the loadings to all compound-concentration combinations to create new composite scores; also we eliminated some individual ticks that seemed to have unusual behaviors (producing an unusual composite score). This did not result in better (or more interpretable) separation, deet at 413 nmol/cm^2 filter paper was still the only one that significantly differed from ethanol and the ranked means did not match their respective concentrations. We also examined the individual behaviors' relation to the compound-concentration combinations using summary statistics and graphics and found no obvious pattern. In all models, R^2 was relatively small (about 10%), indicating that the model explained little of the variation in the composite scores. Whether or not ticks transferred to the filter paper differed significantly between repellent treatments of the highest concentration tested and the control (p = 0.002), but no difference was detected between compounds (p= 0.395). Thus, we conclude that our implementation of this test was not effective to test for compound or concentration differences with *A. americanum*.

Petri Dish Choice Bioassay

Both compounds were avoided at higher doses (Figure 4 gives model based means with a 95% confidence interval), with deet showing some decline in repellency with time (the positive linear time trend was significant; p = 0.031, t-test, 91 d.f.). The wide 95% confidence intervals are due to the relatively small sample sizes used. The over-dispersion parameter was estimated to be only about 1.4 (where 1.0 indicates no over-dispersion). The asymmetry in the confidence intervals is due to the back-transformation, the 95% confidence intervals are symmetric on the logit scale.

Table 3. Tick behaviors used to construct composite scores for moving object bioassay

Behavior	*Loading, 1st Principal Component*
Reach final 1 cm of wire (yes/no)	1.66
Reach final 0.5 cm of wire (yes/no)	-4.29
Contact with paper (yes/no)	2.20
Transfer to paper (yes/no)	0.562
Drop from paper (yes/no)	1.19
Time to final 1 cm of wire	0.000526
Time to tip of wire	-0.0145
Time to transfer to paper	0.00340
Time left paper	-0.0150

Variation and Repeated Use of Ticks

To determine if ticks fatigue (change in response) with continuous testing, we tested each tick twice a day (tested on days 1, 2, 3, 4, 8, 9). The following mixed model was fit to the data (with variance estimates of 0.418 and 3.668 for the among tick and residual components, respectively): $y = -0.442 + 0.378\,x_1 - 0.078\,x_2 + 0.521\,x_3$, where y = composite score, x_1 = day (with values 0, 1, 2, 3, 7, 8, s.e. = 0.0629); $x_2 = (x_1 - \text{mean}(x_1))^2$, s.e. = 0.0263; $x_3 = 0$ for AM and = 1 for PM (i.e. a dummy variable), s.e. = 0.290, p-value estimates for the coefficients of the x variables were 0.000, 0.004, and 0.074, respectively. The regression equation can be interpreted as the composite score generally increasing (ticks exhibiting reduced performance, more easily repelled) as day increases, though with some curvature due to the quadratic component, and with a marginally significant effect of time of day (composite scores generally higher in PM). Tick to tick variation was moderate, but the large residual variance indicates that, for each tick, there was considerable variability in composite score from one trial to the next, as shown in Figure 5 for a few example ticks. These results demonstrate that continuous testing adversely affects tick performance.

To determine if a less intense schedule ameliorated the repeated testing effect, we tested another group of ticks once per day (tested on days 1, 6, 19, 22, 26). The following mixed model was fit to the data (with variance estimates of 0.823 and 2.445 for the among tick and residual components, respectively): $y = -1.179 + 0.067\,x_1$, where y = composite score, x_1 = day (with values 0, 5, 20, 23, 26, s.e. = 0.013), p-value estimate for the coefficient of x_1 was 0.000. A quadratic effect examined in a preliminary model was not significant. While the slope is shallower, the results suggest that there is still a repeated testing effect.

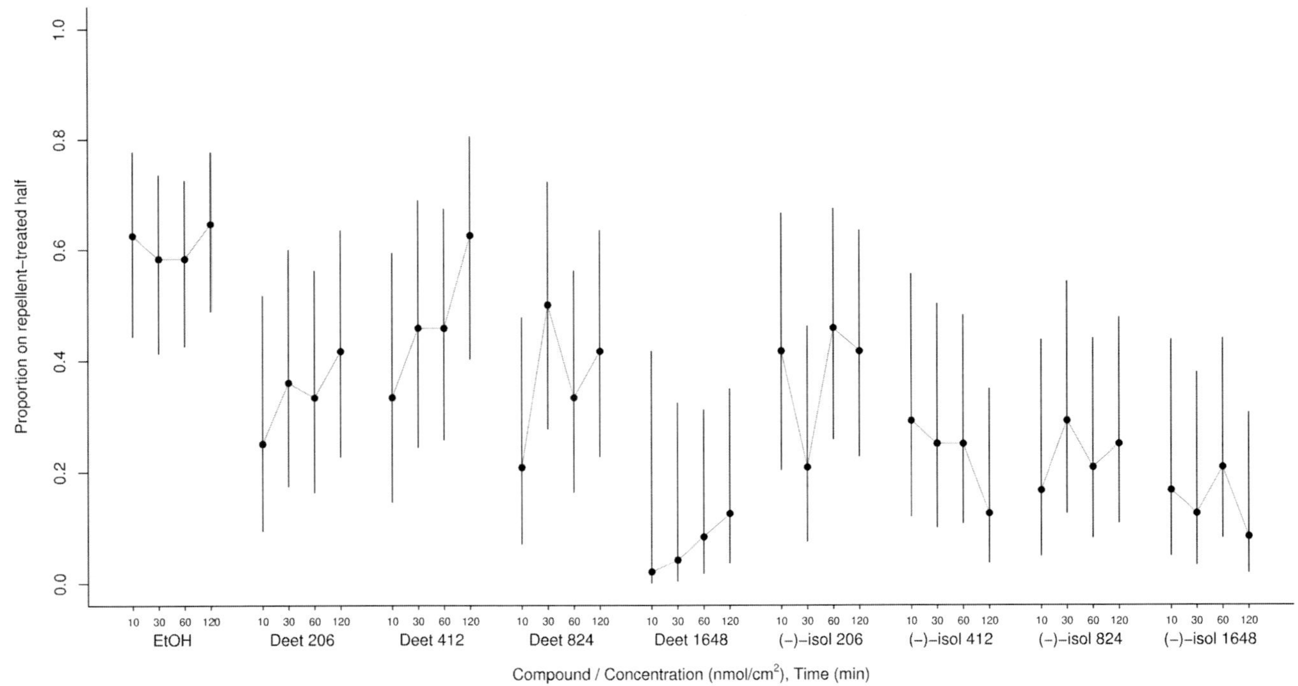

Figure 4. Points give the proportion of ticks on the repellent-treated half of a filter paper in a Petri dish at 10, 30, 60, and 120 min for various concentrations of deet and (-)-isolongifolenone. Vertical bars give 95% confidence intervals (asymmetric on the back-transformed proportion scale).

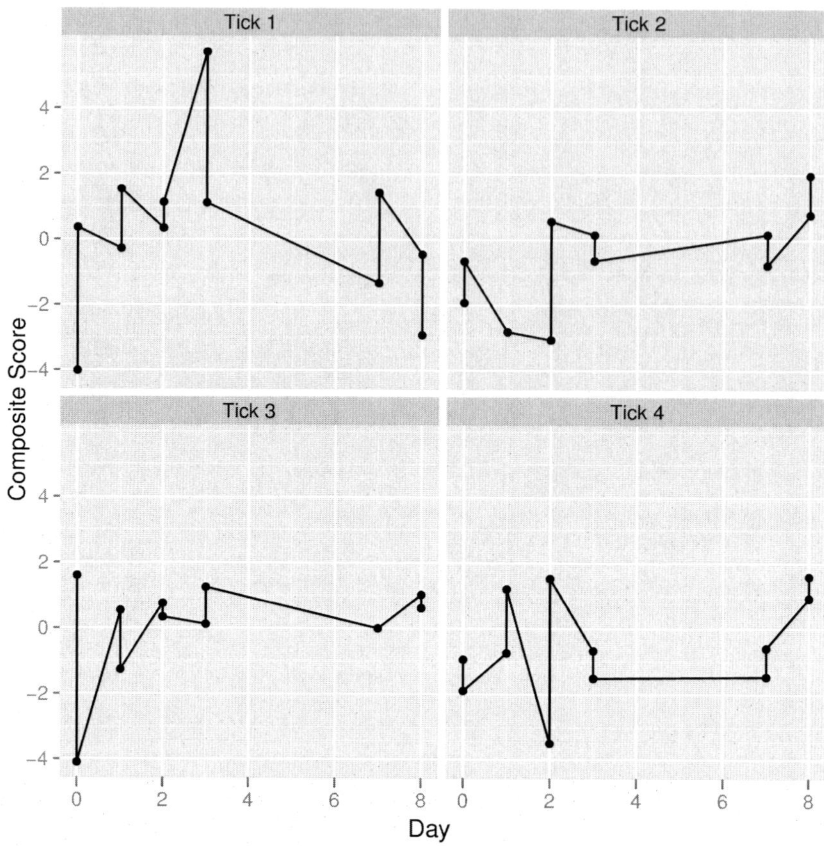

Figure 5. Composite scores over days (ticks tested twice a day) for four example ticks that were repeatedly tested on vertical filter paper. Note that the composite scores are rather erratic over time and the patterns among ticks dissimilar.

An alternative to testing each tick many times is to test it twice. We tested each tick twice on the same day, and analyzed the resulting composite scores in a mixed model using a factor with two levels (AM and PM). This factor was not significant, $p = 0.792$, suggesting that a second test on the same day does not decrease performance if each tick is only tested twice. Variance estimates were 0.000 and 2.861 for the among tick and residual components, respectively.

In our last set of trials we wanted to determine if a 2-wk 'recuperation' time (tested twice on day 1, tested once on days 15 and 16) would ameliorate the repeated testing effect. A factor was created with 4 levels (for the 4 trials per tick). In a mixed model (with variance estimates of 0.000 and 1.748 for the among tick and residual components, respectively), this factor was not significant, $p = 0.777$, suggesting that a two week recuperation time is sufficient for ticks to regain prior performance levels.

Thus, we found that repeatedly testing ticks does not decrease performance, if they are (1) tested twice a day only, and (2) they are allowed to 'recuperate' from the first day of testing for 2 wk before retesting. Thus, researchers can benefit from these results since they show that one can use one half to one fourth as many ticks to produce comparable results. The price paid for this is that the ticks must be held for a 'recuperation' period and that the statistical models used must allow for the correlation induced by repeated testing of the same tick. This correlation was estimated to be zero when testing ticks twice in the same day and when allowing for a 'recuperation' period (and was small in other tests), so that the repeated testing does not greatly affect effective sample size (if the correlation was high, then the effective sample size can be much smaller than the number of ticks actually used).

Discussion

Dautel (*20*) reviewed an array of methods used to assess the efficacy of tick repellents. He grouped the methods in three categories: 1) those using live hosts, 2) those using attractants associated with hosts, and 3) those using no attractants. The *in vitro* bioassays we examined fall into categories 2 and 3, with the petri dish bioassay essentially lacking host cues and the MOB using the simulated host cues, temperature and motion. In the MOB and the other bioassays, ticks are exposed to host cues in the form of chemical, vibrational and visual (*A. americanum* possess eyes) stimuli from an observer/experimenter situated nearby.

We tested the same stage of the same species of tick against mostly the same concentrations of two repellents tested under nearly the identical conditions (solvent, filter paper, temperature and RH range). The similarity in effectiveness between deet and (-)-isolongifolene reported by Zhang et al. (*16*) was confirmed in the various types of bioassays. In certain types of bioassays, higher proportions of *A. americanum* nymphs were repelled (e.g. 4 × 7-cm offset ring configurations) compared with other bioassays (e.g. 4 × 7-cm basic configuration). These findings provide more evidence that a panel of test compounds must include at least one 'standard' repellent, such as deet, to provide a common basis or link for comparing results from bioassays that use different methods.

The tendency of *A. americanum* and other ticks to climb has been used in several *in vitro* and *in vivo* bioassays. In testing fractioned compounds from *Chamaecyparis nootkatensis* (D. Don) Spach. essential oil, Dietrich et al. (*26*) allowed *I. scapularis* to climb a vertical cotton-tipped applicator with test solutions applied to its apical portion. The repellency of benzoquinone compounds secreted defensively by millipedes was tested by releasing *A. americanum* nymphs on a clay substrate, and encircled by a ~3-cm high cylinder of filter paper to which a 2-cm wide band of test solution had been applied (*27*). In fingertip bioassays (*16*, *27–29*) used to evaluate repellent efficacy against *A. americanum*, the finger was held vertically with the untreated tip down. Cream, spray and lotion formulations of repellents were applied in a 5-cm wide ring encircling each ankle of human volunteers and challenged by ticks that were placed or crawled onto the volunteers' feet at 2-h intervals for 12 h (*30*, *31*). A similar test using a treatment on the wrist and forearm is the bioassay recommended by the EPA for obtaining data for registration of tick repellents (*32*).

The results of the 22 × 1-cm vertical filter paper bioassay indicate that in climbing type bioassays with *A. americanum* nymphs the vertical dimension need not be great, with 8-9-cm height optimal. When the strip was treated with ethanol alone, 21 of 54 (38.9%) of the nymphs never climbed the full 20 cm "treated" section in 10 min, with 28.1% of these ticks dropping from the strip. On ethanol treated 22 × 1-cm strips, 94.4% of 54 ticks climbed past 5 cm in 10 min, and 79.6% climbed past 8 cm. In the interest of having robust controls, a vertical treatment of 5 cm may be a good option.

The narrowness (1 cm) of the 22-cm strip allowed little lateral movement by ticks. A critical dimension in barrier type repellent tests is the minimum distance across the treated surface that a tick must traverse to defeat the treatment. A fast moving tick that enters an overly narrow barrier treatment might quickly detect a decreasing gradient of repellent toward the opposite border of the treatment and continue through the barrier, whereas a broader barrier would allow more opportunity for a tick to retreat from or drop off the treated surface. How narrow is too narrow? On the 22 × 1-cm strips treated with 1238 nmol deet or (-)-isolongifolenone/cm^2 filter paper, >50% of the ticks did not climb above 1 cm. As the heights increased, the proportions of ticks not reaching those heights (repelled) increased, so heights approaching 8 cm would give better discrimination from controls.

In the 22 × 1-cm bioassay, the proportions of ticks reaching the height increments 1-10 cm tended to decrease as the concentrations of the two repellents increased to 1238 nmol repellent/cm^2 filter paper, but at the highest concentration (1650 nmol repellent/cm^2 filter paper) the proportions were similar to those for 413 and 825 nmol repellent/cm^2 filter paper. This variability is reminiscent of that observed in the Zhang et al. (*16*) data for (-)-isolongifolenone in fingertip tests with *A. americanum*. Analysis of the 22 × 1-cm strip tests, show that climbing-type bioassays need not last long with ~6 min duration capturing the critical data. Because adequate replication is essential, minimizing the duration of individual bioassays is important.

We thought that the addition of the loop (offset and direct) to the lower untreated area of the 4 × 7-cm filter paper might preempt the "run or drop" behavior of *A. americanum* by providing an alternative escape from the repellent, but such was not the case. The addition of the ring below the 1 cm untreated area at the lower end of the filter paper may have enhanced the likelihood of ticks dropping from the paper. We have used the basic 4 × 7-cm filter paper bioassay in several studies (e.g. (*14*, *15*, *33*)), and in our experience, only rarely ≥2 of 10 ticks fell from untreated controls. In the case of the basic 4 × 7-cm tests we report here, no ticks (n = 30 tested) fell from the controls. However, in 10 of 16 controls in the loop bioassays ≥2 ticks fell, with 0.40 of the ticks (n = 70) falling in the controls of the offset (near) loop bioassay. When the loop was on the same side of the 4 × 7-cm strip as the observer, it allowed ticks to approach as close as ~4 cm to the observer and away from the repellent. With this configuration, the highly active *A. americanum* ticks may have fallen from the loop in an attempt to reach the observer. For the two highest doses of deet and (-)-isolongifolenone (825 and

1650 nmol compound/cm^2 filter paper), higher proportions of ticks were repelled and remained below the treated area in the offset (near) loop bioassays than in the basic 4 × 7 cm bioassays. The apparent higher repellency may be due to a greater tendency of ticks to drop from the loop configurations, as seen in the controls, or to the escape option of the loop, manifested in ticks remaining below the treated area at 10 min. The problem with ticks falling from the paper makes the offset loop (near) bioassay unsatisfactory.

For the MOB, we recorded the same behaviors as Dautel et al. (*19*) who tested deet (0.11 mg/cm^2 filter paper) and ethanol controls against *Ixodes ricinus* (L.). Several important behaviors associated with host acquisition (contacting, transferring to and remaining on a moving surface) were recorded in the MOB, but like Dautel et al. (*19*) we only found significant differences between treatment and control for the proportion of ticks that transferred to the filter paper and the length of time the ticks remained on paper. While the behaviors recorded seemed well suited for the composite score analysis, separation of the treatments did not occur. Testing an additional higher concentration for this and the other bioassays might have improved discrimination. In our tests, at low concentrations of repellent ticks left filter paper treated with (-)-isolongifolenone more quickly than deet-treated paper. Two factors confound "time on paper" results. First, ticks move at different speeds, which is evident in untreated controls. Second, when a tick transfers to the moving filter paper, it is not equidistant to all the edges of the paper; the distance a tick travels to the edge of the paper depends on where it gets on the paper and the direction it crawls. With highly effective concentrations of repellent, *A. americanum* would be expected to quickly fall from the paper, which we observed.

The petri dish choice bioassay differs from the other bioassays in this study in that it offered no opportunity for a tick to remove itself more than 4.5 cm from the repellent treatment. Larger petri dishes would allow ticks to escape further from the repellent. With an active tick like *A. americanum*, the absence of a complete escape option may create a stronger challenge to the repellent than vertical tests, but probably not as strong a challenge as placing the ticks within a horizontal ring of a repellent, as used by Carroll et al. (*14*). The no escape feature in petri dish tests has been used to force ticks to choose to contact (or avoid) either of two repellent treatments, when each half of the substrate received a different repellent treatment (*34*). Once set up, the petri dish bioassay does not require the constant attention of an observer. We recorded tick locations periodically for 2 h after placing the ticks between the filter papers, which allowed detection of a decline in efficacy of deet over time.

The highest concentration we used (1650 nmol compound/cm^2 filter paper) was rather effective, but even 3 times that concentration in 4 × 7-cm vertical filter paper bioassays under the same conditions does not repel all *A. americanum* nymphs (Carroll, unpublished data). The dose range used in these tests may have been adequate, but the addition of a higher (even more repellent) concentration would have given a more complete picture of dose response relationships. The highest concentration of (-)-isolongifolenone repelled 90% of *A. americanum* in the 4 × 7-cm strip in the offset ring (near) configuration, the highest concentration of deet repelled 86.7% of the ticks in the 4 × 7-cm strip in the offset ring (opposite) configuration, and in the petri dish bioassay no ticks were in the treated half at 10 min.

Because rearing or purchasing ticks for laboratory use can be costly, reusing individual ticks in bioassays is a reasonable option if a similarity in response between reused ticks and naïve ticks could be assured. By retesting individually tracked *A. americanum* nymphs, we found that the ticks' performance was negatively affected by repeated testing. While testing the same ticks twice in one day (morning and afternoon) did not produce different results, based on our findings continued once or twice daily testing is inadvisable. Ticks allowed 2 wk to recover between testing responded similarly to naïve ticks. Eventually age related changes in tick activity and responses can be expected.

Although there are advantages to using *A. americanum* in repellent bioassays, some drawbacks exist. Perhaps, the greatest challenge in using *A. americanum* in behavioral bioassays involves transferring these particularly active and tenacious ticks into vials or bioassay arenas. In this regard, pump operated aspirators are useful in capturing ticks and putting them in vials. None of the climbing bioassays used in this study seemed to mitigate the variation observed in Zhang et al. (*16*). Further investigation is needed for a better understanding of the nuances of *A. americanum* responses to repellents.

Acknowledgments

We thank James McCrary and Abdul Saboor Khan, USDA, ARS, Invasive Insects Biocontrol and Behavior Laboratory, Beltsville, MD for conducting bioassays, Jun Nie, USDA, ARS, Invasive Insects Biocontrol and Behavior Laboratory, Beltsville, MD for preparing test solutions, and the USDA, ARS, Knipling-Bushland U.S. Livestock Insects Research Laboratory, Kerrville, TX for providing *A.americanum* nymphs.

This article reports the results of research only. Mention of trade names or commercial products in this article is solely for the purpose of providing specific information and does not imply recommendation or endorsement by the U.S. Department of Agriculture for it use.

References

1. Parola, P.; Raoult, D. D. *Clin. Infect. Dis.* **2001**, *32*, 897–928.
2. Piesman, J.; Eisen, L. *Annu. Rev. Entomol.* **2008**, *53*, 323–343.
3. *Lyme Disease*; Department of Health and Human Services, Centers for Disease Control and Prevention: Ft. Collins, CO, 2002.
4. Bissinger, B. W.; Roe, R. M. *Pestic. Biochem. Physiol.* **2010**, *96*, 63–79.
5. Syed, Z.; Pelletier, J.; Flounders, E.; Chitolina, R. F.; Leal, W. S. *PLoS ONE* **2011**, *6* (3), e17705, DOI: 1371/journal.pone.0017705
6. Childs, J. E.; Paddock, C. D. *Annu. Rev. Entomol.* **2003**, *48*, 307–337.
7. Stromdahl, E. Y.; Vince, M. A.; Billingsley, P. M.; Dobbs, N. A.; Williamson, P. C. *Vector-Borne Zoon. Dis.* **2008**, *8*, 15–24.
8. Keirans, J. E.; Durden, L. A. *J. Med. Entomol.* **1998**, *35*, 489–495.

9. Ginsberg, H. S.; Ewing, C. P.; O'Connell, A. F., Jr.; Bosler, E. M.; Daley, J. G.; Sayre, M. W. *J. Parasitol.* **1991**, *77*, 493–495.
10. Means, R. G.; D. J. White, D. J. *J. Vector Ecol.* **1997**, *22*, 133–145.
11. Waladde, S. M.; Rice, M. J. The Sensory Basis of Tick Feeding Behavior. In *Physiology of Ticks*; Obenchain, F. D., Galun, R., Eds; Pergamon: New York, 1982; pp 71−118.
12. Wilson, J. G.; Kinzer, D. R.; Sauer, J. R.; Hair, J. A. *J. Med. Entomol.* **1972**, *9*, 245–252.
13. Carroll, J. F.; Solberg, V. B.; Klun, J. A.; Kramer, M.; Debboun, M. *J. Med. Entomol.* **2004**, *41*, 249–254.
14. Carroll, J. F.; Klun, J. A.; Kramer, M. *J. Entomol. Sci.* **2008**, *43*, 426–430.
15. Carroll, J. F.; Paluch, G.; Coats, J.; Kramer, M. *Exp. Appl. Acarol.* **2010**, *51*, 383–392.
16. Zhang, A.; Klun, J. A.; Wang, S.; Carroll, J. F.; Debboun, M. *J. Med. Entomol.* **2009**, *46*, 100–106.
17. Da Silva, T. B. C.; Alves, V. L.; Mendonca, L. V. H.; Conserva, L. M.; Da Rocha, E. M. M.; Andrade, E. H. A.; Lemos, R. P. L. *Pharm. Biol.* **2004**, *42* (94−97), 517.
18. Kramer, M.; Weldon, P. J; J. F. Carroll, J. F. *Anim. Behav.* **2009**, *77*, 763–768.
19. Dautel, H.; Kahl, O.; Siems, K.; Oppenrieder, M.; Müller-Kuhrt, L.; Hilker, M. *Entomol. Exp. Appl.* **1999**, *91*, 431–441.
20. Dautel, H. *Int. J. Med. Microbiol.* **2004**, *293* (Suppl. 37), 182–188.
21. Dautel, H.; Cranna, R. *Aust. Vet. Pract.* **2006**, *36*, 138–147.
22. Douglas, H. D., III; Co, J. E.; Jones, T. H.; Conner, W. E. *J. Chem. Ecol.* **2004**, *30*, 1921–1935.
23. Crystal, M. M.; Demilo, A. B. *J. Georgia Entomol. Soc.* **1984**, *19*, 517–523.
24. Weldon, P. J.; Carroll, J. F.; Kramer, M.; Bedoukian, R. H.; Coleman, R. E.; Bernier, U. R. *J. Chem. Ecol.* **2011**, *37*, 348–359.
25. Pinheiro, J.; Bates, D.; DebRoy, S.; Sarkar, D. *Linear and Nonlinear Mixed Effects Models*, R package version 3.1-101; R Development Core Team, 2011.
26. Dietrich, G.; Dolan, M. C.; Peralta-Cruz, J.; Schmidt, J.; Piesman, J.; Eisen, R. J.; Karchesy, J. J. *J. Med. Entomol.* **2006**, *43*, 957–961.
27. Carroll, J. F.; Kramer, M.; Weldon, P, J.; Robbins, R. G. *J. Chem. Ecol.* **2005**, *31*, 63–75.
28. Schreck, C. E.; Fish, D.; McGovern, T. P. *J. Am. Mosq. Control Assoc.* **1995**, *11*, 136–140.
29. Carroll, J. F.; Cantrell, C. L.; Klun, J. A.; Kramer, M. *Exp. Appl. Acarol.* **2007**, *41*, 215–224.
30. Carroll, J. F.; Benante, J. P.; Klun, J. A.; White, C. E.; Debboun, M.; Pound, J. M.; Dheranetra, W. *Med. Vet. Entomol.* **2008**, *22*, 144–151.
31. Carroll, J. F.; Benante, J. P.; Kramer, M.; Lohmeyer, L. H.; Lawrence, K. *J. Med. Entomol.* **2010**, *47*, 699–704.
32. Product Performance Test Guidelines, Insect Repellents To Be Applied to Human Skin; OPPTS 810.3700; U.S. Environmental Protection Agency: Washington, DC, 1999.

33. Carroll, J. F.; Tabanca, N.; Kramer, M.; Elejalde, N. M.; Wedge, D.; Bernier, U. R.; Coy, M.; Becnel, J. J.; Demirci, B.; Başer, K. H. C.; Zhang, J.; Zhang, S. *J. Vector Ecol.* **2011**(in press).
34. Bissinger, B. W.; Apperson, C. S.; Sonenshine, D. E.; Watson, D. W.; Roe, R. M. *Exp. Appl. Acarol.* **2009**, *48*, 239–250.

Chapter 8

Development of Space Repellents for Vector Control

J. P. Grieco[*] and N. L. Achee

Uniformed Services University of the Health Sciences, Bethesda, Maryland 20814, U.S.A.
[*]E-mail: jgrieco@usuhs.mil

Arthropod-borne diseases impact large portions of the developing world and impart substantial economic and health burdens in these regions. Despite the burden these afflictions have on local populations, our tools for controlling the vectors responsible for pathogen transmission are limited. One critical component of any vector-borne disease management strategy is the use of chemicals in either indoor residual sprays, on bed nets or as topically applied repellents. The chemicals that are currently recommended for use, however, are quickly becoming inadequate to sustain disease control due in part to insecticide resistance. Evaluation of how mosquitoes respond to insecticides is an expanding field of study and the knowledge gained from these endeavors is paramount to advancing the development of new classes of chemistry to expand our current arsenal of effective compounds. One area of particular interest is the exploitation of behavior-modifying actions of chemicals in order to create vector free spaces and thereby reduce human-vector contact. Such chemicals could be used in various delivery platforms and in combination with other vector control interventions to enhance the effectiveness, affordability and sustainability of public health tools.

Alternate Actions of Insecticides

Knowledge of how mosquitoes respond to insecticides is of paramount importance in understanding how an insecticide functions to prevent pathogen transmission. In order to truly understand how chemicals function to break pathogen transmission one must first determine the primary mode of action for a chemical by ordering the activity based on the concentration at which the specific response is elicited. Chemicals exert different actions on mosquitoes to include contact irritancy, spatial repellency and toxicity (*1*). Evidence suggests that the behavioral response to spatial repellent and contact irritant actions are separate (or independent) from the toxic action of a compound. Various degrees of activity can occur even within the same chemical class. Laboratory and field assays also show chemicals like pyrethroids and DDT induce behavior-modifying actions, such as contact irritancy and spatial repellency, at concentrations far below toxic levels (*2*). In addition, these actions will continue to be exhibited in insecticide-resistant test populations. This information facilitates new uses of our existing arsenal of insecticides for the control of vector-borne diseases. The search for more effective active ingredients and novel delivery platforms will only increase as behavior modifying actions continue to show success in reducing the burden of disease (*3*).

The current emphasis for vector control is to label almost any chemical used against insects as an "insecticide." By definition, an insecticide (insect-icide or insect-icidal) is a chemical that is used to kill insects. This single term does not adequately address in a meaningful way the various toxic and behavioral ways in which these chemicals impact on the vector population. However, throughout history this single term has been the foundation for the old paradigm that classifies chemicals sprayed in efforts to control vector-borne disease based solely on a killing action. Although actions outside of toxicity have been recognized for decades, it is only recently that research has focused on systematic laboratory and field studies to quantify these other actions and their potential impact on pathogen transmission to account for how the behavioral actions of these chemicals result in disrupting man-vector contact and thereby breaking pathogen transmission. These repellent and irritant actions were first documented more than 60 years ago (*4*) but they were never broadly perceived as being important, illustrating how lack of appropriate labels and a conceptual framework of multiple chemical actions can work against a clear understanding of how these chemicals work in the field. Today, any discussions about insecticides for vector control operate under the assumption that the chemical is toxic and that it only functions by killing mosquitoes. Over 45 years ago Dethier (*5*) showed that chemicals elicit multiple actions and that insects respond to those actions through a variety of behavioral responses. He noted that if we were to take a closer look at modes of action, we could find a much more diverse set of terms for movements of insects toward or away from a chemical source. As early as 1953, Muirhead-Thomson (*6*) concluded chemicals could disrupt contact between humans and malaria-transmitting mosquitoes and stop pathogen transmission without killing the mosquitoes. Subsequent authors speculated that spatial repellents applied to house walls could have advantages over topical repellents that are traditionally applied to the skin. In contrast to topical repellents, repellents designed for application to the interior walls of houses could be formulated to have a longer residual life that could result in reduced frequency of application and production costs. Regardless, the search for alternative compounds has focused almost entirely on toxicity. Insecticides recommended for IRS continue to be evaluated almost entirely on mosquito mortality (*7*) and laboratory evaluations continue to use toxicity as the primary measure of success (*8–10*).

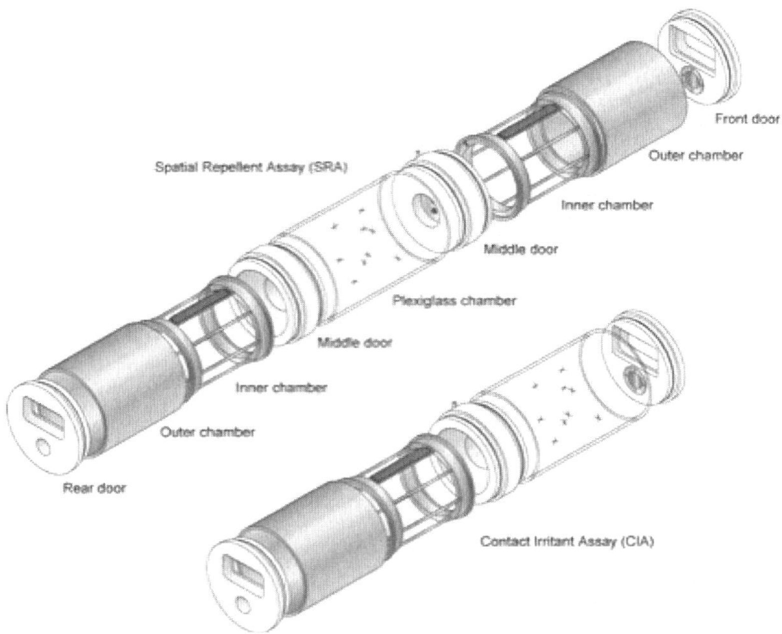

Figure 1. High throughput assay (1) system for the evaluation of repellent and irritant actions elicited by chemicals.

Assays have been developed to examine insects behavioral responses to chemicals in a more systematic and repeatable fashion (*1*, *2*) (Figure 1). In order to measure chemical actions one must first define what actions are of interest from an epidemiological standpoint and establish the entomological endpoints (i.e. the insects response to the chemical) that correlate with reduction in pathogen transmission. A toxic action produces knockdown or death after the mosquito makes contact with the chemical in either the solid or vapor phase. A contact irritant action stimulates movement away from the chemical source after the mosquito makes physical contact. A spatial repellent action stimulates movement away from the chemical source without the mosquito making physical contact with the treated surface.

In order to accurately measure the spatial repellent action, a weighted spatial activity index (WSAI) was used to quantify repellent induced movement and is based upon the oviposition activity index of Kramer and Mulla (*11*), was used to evaluate the responses of female mosquitoes in the spatial repellency assay. The WSAI measures the proportion of females in the control chamber over the treated chamber after correcting for the proportion of females in the control chamber multiplied by the number of test subjects that responded in either direction. Negative WSAI values indicate an attractant response, positive values indicate a repellent response and values of 0 or close to zero indicating no response. In other words, a WSAI value of -20 would indicate that a greater proportion of mosquitoes moved into the treatment chamber than the control chamber thus indicating an attractant response. A WSAI value of 20 would indicate that a greater proportion of mosquitoes moved into the control chamber (away from the treatment end of the assay device) indicating a repellent action.

Thresholds exist for when and how insects respond to chemical actions. The thresholds are governed by intrinsic and extrinsic factors such as inherent strength of a chemical action, chemical volatility, ambient temperature, humidity, proximity and length of exposure, and a mosquito's physiological status and sensitivity to a compound, to name just a few factors. The concentration dependent order in which thresholds are exceeded determines whether the primary mode of chemical action is repellent, irritant or toxicant under ambient environmental conditions. Research has documented that house wall residues of three important and commonly used insecticides elicit varying combinations of behavioral actions. Based on results from laboratory tests and experimental hut studies, criteria were established for revising classifications of chemicals that are presently recommended for use in malaria control programs. This revised classification scheme proposes a new paradigm that emphasizes a single or the combination of multiple chemical actions to control pathogen transmission by breaking human-vector contact (*12*). This new paradigm will permit members of the public health community to discuss and characterize chemicals according to their true modes of action.

A probability model has been developed to highlight this new paradigm (Figure 2) which looks at the composite impact of a chemical used inside a structure. This model assumes that a hundred mosquitoes would enter a house, bite while indoors, and escape and survive if the house were not sprayed (*13*). Through a series of experimental huts studies we have been able to assign actual values to the various model parameters to establish the ultimate impact of a chemical which elicits irritancy, repellency and toxicity.

In huts sprayed with DDT, 59 of the 100 mosquitoes would not enter (*12*). Of the 41 that enter, 2 would die and fall to the floor. Of the 39 survivors, 12 would exit prematurely. One of the 12 mosquitoes that escaped would die within the next 24 hours. This leaves 27 mosquitoes that theoretically could bite and survive. However, it is important to understand that chemical is present in houses 24 hours each day, these statistics cover only 7 hours, not 24. These statistics suggest that DDT reduced risk from 100 mosquitoes by 73% within the first 7 hours.

In huts sprayed with alpha-cypermethrin, all 100 mosquitoes would enter the house (*12*). Of the 100 that entered, 15 would die. Of the remaining 85, 46 would exit prematurely and 9 of those would die. This leaves 39 mosquitoes that theoretically could bite and survive. The spatial repellent, contact irritant, and toxic actions of alpha-cypermethrin sum to 61% protection.

In the case of malaria control, success through the mechanism of spatial repellency means that a chemical functions as a form of chemical screening, which discourages mosquitoes from entering houses and thus interupts the human-vector contact at a critical point: when people are sleeping in their homes. The new classification scheme that we have proposed, characterizes chemicals on the basis of spatial repellent, contact irritant and toxic actions. The first criterion for evaluating a chemical is the concentration at which the chemical exceeds a threshold for vector response. If mosquitoes are intoxicated at concentrations lower than that required for a behavioral response then toxicity supersedes other actions since the insect might be overcome before being stimulated through mechanisms of contact irritancy or spatial repellency. Likewise, if an irritant response occurs at a lower concentration of chemical than required for toxicity, then the irritant response precludes toxicity since the insect or some proportion of insects may move away from the chemical before acquiring a lethal dose. These relationships are even more pronounced for a spatial repellent action. If a spatial repellent response is stimulated by a lower or equal concentration of chemical than required for either contact irritancy or toxicity, then the insect or some proportion of insects will be repelled without making contact with the chemical. Thus the three chemical actions (spatial repellency, contact irritancy, toxicity) can be quantified according to proportional dose-response relationships, and relative rank order of actions can be defined.

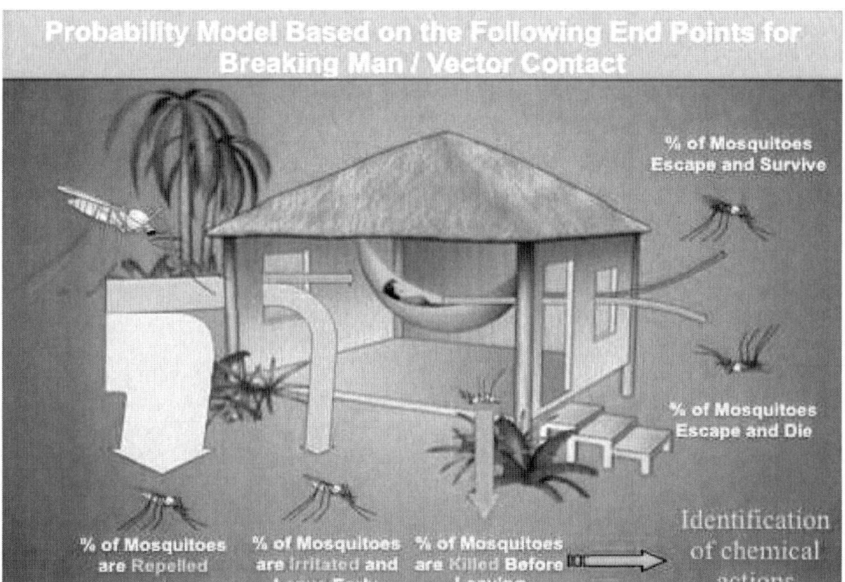

Figure 2. Probability model (13) that describes the various actions of chemicals applied in a home and the epidemiological impact for breaking human vector contact.

There is considerable debate as to the role of irritancy in this framework. If the role of a repellent is to be a chemical screen and create a protective barrier around a house, how would one describe a compound that puts the mosquito in an excited state and forces it from the home. In this case, the mosquito still enters the house and could potential bite and transmit disease. So from an epidemiological perspective there is the issue of whether a contact irritant would truly be beneficial. Another potential drawback of an irritant is the requirement that a mosquito must first make tarsal contact with the treated surface. In any treated home there will always be competing untreated surfaces in the house that may further decrease the effectiveness of an irritant or toxic compound. If an irritant compound has the combined effect of being toxic, as many currently labeled vector control chemicals are, this action may decrease the impact of the toxic action of the compound. In other words, irritancy could reduce the contact time with a treated surface, thus reducing the potential for picking up a lethal dose of the compound. This was demonstrated in hut studies when using a compound like alpha-cypermethrin that functions as both an irritant and a toxicant. Mosquitoes were collected from exit traps after being exposed to this chemical and were held for 24 hours to determine mortality rates. Data clearly showed that alpha-cypermethrin caused a very rapid exiting response in the population as compared to control. The more interesting result, however, was mortality rates in the escaping populations were very low (less than 30%). These findings were also demonstrated with other highly irritant compounds like deltamethrin and permethrin. More work must be performed to better understand the true role of irritancy and its benefits and impediments in disease control.

Current Research Efforts on Spatial Repellents

The focus on the use of spatial repellents in the field has increased dramatically in recent years as the idea of disease reduction through spatial repellency gained credibility. Several lines of research are being explored to take advantage of the unique potential that spatial repellents offer over traditional thoughts on IRS or use of long-lasting insecticide treated nets (LLIN).

Treated Materials

The practice of IRS has become increasingly problematic as it requires trained personnel and the ability to treat vast numbers of houses. The use of LLINs are aimed at individual protection of those directly under the net, however, community wide protection has been reported if a sufficient level of coverage is achieved. The scientific community is currently struggling with several issues associated with bed nets, not the least of which is the technology relies heavily on pyrethroid compounds which are quickly losing their effectiveness due to insecticide resistance. The concept of using treated materials in a spatial repellent or contact irritant strategy could include the focal placement of repellent materials in portals of entry into the house or the placement of irritant material in indoor locations that are preferred resting sites to make the interior of the house less acceptable for vector populations. This strategy minimizes the amount of active ingredient needed for treatment of an entire house and creates a consumer product platform that transfers ownership of the personal protection to the individual. In theory, personal ownership will make the method more sustainable.

Many homes around the world currently make use of a variety of decorative ribbons and banners for beautification (Figure 3 A). By taking advantage of such household items, an experimental study was carried out to determine if the same material could be treated with a spatial repellent, and be placed in portals of entry into a house to reduce mosquito entry. The study was carried out in an experimental hut design in Belize, Central America. The evaluation consisted of treating three different pieces of ribbon with either a spatial repellent compound (Compound A) or one of two pyrethroids (Compounds B and C) at the recommended field application rate labeled for each compound (*14*, *15*). The material treated was a 1.5-cm polyester ribbon purchased locally. Once treated the ribbons were positioned in the center of a 30-cm eave gap of three different experimental huts (Figure 3 B). A matched control was also run that contained a similar ribbon with no chemical application. Collections were conducted from the inside of all four huts over a series of three consecutive nights. Results showed that the repellent compound excluded greater than 80% of the mosquitoes from entering the hut whereas the pyrethroids did little to reduce entering populations (Table 1). This was not surprising because studies showed the pyrethroids used in this study functioned primarily as irritant compounds rather than having any spatial repellent characteristics. Based on the comparative statistics on treated surfaces for IRS versus ribbons, the treated ribbon requiring over 100 times less active ingredient on a ribbon to achieve an 81% reduction in the biting pressure in a house. These types of focally placed repellent materials and wall linings warrant greater attention to maximize a novel approach to breaking human-vector contract.

Figure 3. (A) decorative ribbons placed in the interior of homes in Thailand. (B) – focally placed repellent-treated ribbon in experimental huts to evaluate the effectiveness of a focal spatial repellent treatment for reducing entry of host seeking mosquitoes.

The results in Table 1 demonstrate how effective a properly positioned repellent can be at disrupting mosquito entry into a home. Data clearly show that there was an 81% decrease in entering *Anopheles vestitipennis* populations at the repellent treated hut (Compound A) compared to the untreated control hut. The two huts treated in the same manner with two different pyrethroid compounds (Compound B and C) each showed no significant difference in entry compared to control.

Table 1. Results from a focally applied spatial repellent study in which the material was applied to a 1.5 cm strip of polyester ribbon. Chemical A represents a repellent compound while Chemicals B and C represent pyrethroids commonly used in vector control.

Chemical	Conc. of a.i. (g/m^2)	Night 1 T/C[1] (% reduction)[2]	Night 2 T/C (% reduction)	Night 3 T/C (% reduction)	Total T/C (% reduction)
A	2.0	97 / 681 (86%)	135 / 821 (84%)	148 / 526 (72%)	380 / 2028 **(81%)**
B	0.03	663 / 631 (5%)	622 / 631 (2%)	562 / 607 (7%)	1847 / 1869 (1%)
C	0.5	373 / 408 (9%)	561 / 520 (9%)	642 / 623 (3%)	1576 / 1551 (2%)

[1] T/C = treatment (T) and control (C) huts.
[2] % reduction = density compared to a matched control.

This was not surprising given previous studies showed that the pyrethroids used in this study functioned primarily as irritant compounds rather than having any spatial repellent characteristics. The mosquitoes in this situation could simply avoid the treated surface and make entry into the house.

"Push-Pull" Strategies

One such strategy currently being investigated is the concept of a "Push-Pull" approach to vector control. The term "Push-Pull" in this context refers to the use of both a behavior-modifying compound to "Push" vectors from the interior of a home, thus creating a vector-free space, while at the same time employing a peridomestic trap to 'Pull" those vectors that are repelled out of the environment. A number of studies are looking at the potential impact of incorporating several strategies to strengthen the impact of a spatial repellent. Studies being conducted in Peru and Thailand have focused on *Aedes aegypti*. These studies are evaluating the impact of focally placed treated material to make the interior of a home as unacceptable to entering and resting mosquitoes and thereby keep them out of the house. This creates a protective barrier at the house level. Once mosquitoes are excluded to the outside of the house their impact is further reduced by using an outdoor trap that draws them in and kills the repelled population. Such control strategies are of particular interest when dealing with vectors that bite during times when people can be found outside the house such as with crepuscular species that bite at dawn or dusk when inhabitants my be sitting or cooking outside the house. It is also a potential method of control for species like *Aedes aegypti* that bite throughout the day. Success of the push-pull strategy will depend on providing adequate levels of household protection and simultaneously achieving and effective level of mosquito mortality in the peridomestic environment.

Point-Source Emitters

A considerable research effort is also being placed on combining spatial repellents into point-source emitters that would infuse an active ingredient into an air column yet maintain a small profile thus insuring that repellency is the true mode of action rather than contact irritancy. Such emitters or eminators come in the form of mosquito coils, vaporizing mats and fans. The latter two formats rely on an electrical source that is not practical for the developing world in which no electricity or batteries are available. Mosquito coils are popular in many areas of the world as a viable consumer-driven intervention. These devices result in decreased biting and in some cases reduced pathogen transmission (*16–19*). These products, however, are being sold using claims that they are repelling mosquitoes from either an outdoor or indoor space and are based on evaluations with the measurable endpoint of reduction in biting. Reduction-in-biting indices, however, can be the result of a number of behaviors. Low bite numbers may result from toxicity in which the mosquitoes are quickly knocked down and are unable to bite, or irritancy in which the mosquitoes enter an agitated state and escape the treated area before biting, or it could be true repellency. Each of these situations holds a different impact from an epidemiological standpoint. Therefore, measuring the primary modes of action of these interventions is critical although admittedly challenging to measure. Emitters remove some confounding in that they minimize the potential for contact irritancy behavior. While this is beneficial, it makes measuring the other variables more difficult. As more chemical is volitalized there is an increase in the possibility of knockdown occurring in the vapor phase.

Any chemical will establish a concentration gradient with greatest concentration proximal to the source. With some chemicals, particularly the pyrethriods, the concentration will eventually approach toxic levels resulting in knockdown of the insects prior to reaching its source. The challenge is to determine where along this gradient the various actions and responses occur and to develop entomological measurements that allow for a finer degree of resolution other than anti-biting. To gain further information, it would be ideal to sample the amount of chemical in the air at points where each response occurs.

Another possibility is that at very low concentrations some chemicals will invoke a switch in response from repellency to attractancy. There is evidence in the literature that compounds that show toxicity and irritancy at high doses also function as attractants at concentrations fall below a particular level. One such example is deet which is toxic at very high dose, irritant at medium doses and attractant at low doses. This presents a serious problem as some of these interventions degrade over time and results in less material in the air column that could potentially result in greater numbers of mosquitoes at the treatment source.

Challenges and Research Needs

To date, a truly efficacious DDT replacement has not been found, and one may never be found without examination of the full characterization of behavioral and toxic actions in which DDT functions. Success through the mechanism of spatial repellency means that DDT functions as a form of chemical screening, which stops mosquitoes from entering houses and thus breaks the man/vector contact at its most critical point: when people are sleeping in their homes. DDT's secondary action stimulates those mosquitoes that do enter to prematurely exit, potentially without biting and transmitting disease. Toxicity is only a third-order action of DDT. For these reasons DDT should be considered a very poor killing agent. We propose that a search for a DDT replacement should focus on the criteria of spatial repellency, contact irritancy and toxicity. A replacement also faces the challenge of being inexpensive and persistent. For a DDT replacement to be found, there must be an investment by both industry and research partners to adopt this new paradigm. Industry must open its chemical libraries to screen compounds on properties other than toxicity (the current standard). Currently, there are 15 compounds recommended by WHO for vector control, representing only four chemical classes (*14*, *15*). The last chemical class to be added occurred in 1990. Twenty years later the global community is continuing to place expectations of protection at the population level using the same four classes of active ingredients. This limited arsenal of adult vector control tools is becoming insufficient to sustain a reduction in disease burden in many disease-endemic countries due mostly to insecticide resistance (*20*). Without industry buy-in, we will continue to be limited in our search for active ingredients, and our toolbox of vector control chemicals will continue to shrink as pyrethroid resistance grows.

The research community must also make strides to take a critical look at spatial repellency and its impact on both entomological and epidemiological end points. A larger data set of evidence must be established to demonstrate disease risk reduction using only excitatory and or repellent properties of residual chemicals (other than DDT). This represents a catch-22, in that all residual chemicals have been screened for toxicity. Thus, toxicity is probably their primary mode of action. Given the methodologies of development with emphasis on toxic actions, repellent actions might always be secondary to toxic actions of newly developed active ingrediants. Without the investment by industry to screen for repellent compounds the chemicals available for proof-of-principle studies are extremely limited. This forces researchers to look at non-residual chemicals and other delivery platforms such as coils and eminators to expand the number of products available for testing the potential of disease reduction by non-toxic means.

Thresholds

Thresholds are involved in several challenges to the advancement of spatial repellency for disease control. These issues are associated with the transmission potential by the vector and characteristics of the chemical dispersion.

One question to be answered is, what level do you need to reduce the biting pressure through repellency to break pathogen transmission. To a large extent the answer is dependent on the vector and disease of interest. A poor vector of malaria may not require a high reduction through repellent means to break disease-transmission. On the other hand, a vector such as *Aedes aegypti* in a dengue-transmission setting may require greater than a 95% reduction to stop dengue transmission. Mathematical models are being developed to address these issues

Another question is what concentration of active ingredient in a volume of air space surrounding a treated surface do you get a desired behavioral response? As you move away from a treated surface the concentration of volatilized material decreases. It becomes critical to know at what concentration or range of concentrations one would find a particular behavioral response. This would determine the protective distance or space. If the amount of chemical can be detected at considerable distances from the point source it might be possible to treat every other house and achieve community level protection. This usage would reduce the amount of chemical needed and the time and manpower required to treat. As mentioned previously, some chemicals vary in their actions based on the concentration. Some chemicals function as attractants at low doses yet elicit toxic or irritant actions at higher doses. This makes it imperative to determine how these chemicals disperse and degrade over time. Linking chemical thresholds with mosquito behavior would help determine when and where a chemical is effective and what behavioral response to expect.

Resistance

The mechanism of action for how repellents function is still unclear. The receptor cites and pathways associated with the physiological changes that occur in the insect to elicit a behavioral response are not well known, but major strides are being made in this direction (*21*). Until some of these questions are answered it will be difficult to determine how resistance mechanisms function to alter the behavioral response if at all. We do have preliminary evidence from work performed in the laboratory with resistant strains of mosquitoes to suggest that the actions of spatial repellent and irritant compounds may be tempered in resistant populations.

An evaluation was conducted in which three irritant compounds (permethrin, alpha-cypermethrin and cypermethrin) applied at their recommended field application rate were tested against three strains of *Ae. aegypti* that were classified as either resistant, tolerant or susceptible by bottle bioassay. Three separate populations for each resistance status were run through the contact irritant assay to determine if there was a detectable change in behavior. Results (Figure 4A and B) demonstrated that the resistant population had the lowest level of behavior compared to control. The strain classified as tolerant showed a higher level of behavioral activity but it was still reduced as compared to the results from the susceptible population. This pattern of reduced behavior was consistent and reproducible across three replicates (A, B and C) for each population. This suggests that the contact irritant response is affected by the resistance status of the mosquito.

When the study was conducted to evaluate the impact on spatial repellency the results were not immediately clear. Two repellents were chosen, transfluthrin and DDT. Transfluthrin showed the same pattern of reduced behavioral response in insects that cooresponded with a higher level of resistance. DDT, on the other hand, showed no change in behavior with increasing resistance to the chemical suggesting that the mechanism of action for this behavior is different than for the pyrethroids. This may account for why DDT continues to have an impact on pathogen transmission despite high levels of resistance in the field (*22*). Yet another instance that demonstrates the uniqueness of DDT.

Despite the fact that there was a decrease in both repellent and irritant responses to certain chemicals as the resistance status changed, in all cases there was still a significant response as compared to control. The difference could be seen in comparing the intensity of the reaction between the different populations that were exposed to the chemical. Therefore, even though the response was dramatically reduced, it still could be quantified and could potentially add to the impact of the compound. This area of research requires a greater understanding and considerable more work to determine the pathways that govern these behavioral responses and how they are altered by physiological conditions such as insecticide resistance, nutritional state, age and even whether infected populations behave differently from uninfected populations.

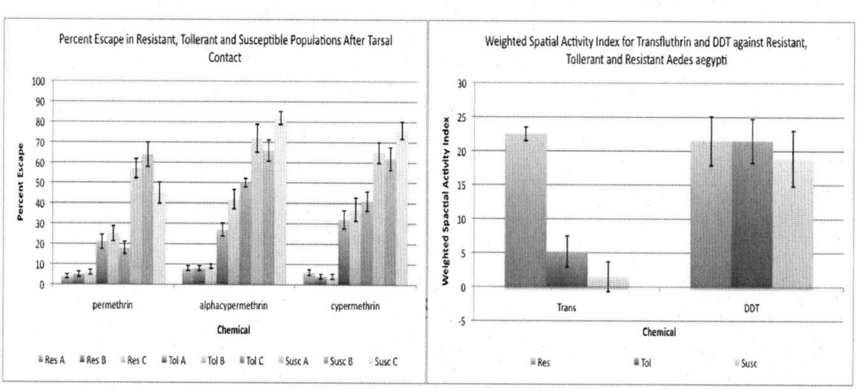

Figure 4. (A) the contact irritant response presented in percent escaping in three populations (Res=resistant, Tol=tolerant, Susc=susceptible) of Ae. aegypti expressing different levels of resistance replicated three times (A, B and C). (B) spatial repellent response to permethrin, alpha-cypermethrin and cypermethrin. Spatial repellent response presented as the weighted spatial activity index for three populations (Res=resistant, Tol=tolerant, Susc=susceptible) of Ae. aegypti expressing different levels of resistance to transfluthrin and DDT.

Diversion

Diversion is defined in relation to spatial repellents as the displacement of mosquitoes from a treated location to an untreated location. Theoretically, diversion could result in a dangerous situation in areas without complete coverage of the spatial repellent. Chemically repelled mosquitoes would aggregate at an untreated location thereby increasing the biting pressure and ultimately increasing the potential for disease. There is evidence of this occurring with topical repellents. If an untreated arm is offered alone, there is a finite, specific biting pressure. If a repellent arm is offered alone there is a marked decrease in the number of mosquitoes that land and feed. When given a choice of two arms with one treated and the other left untreated, the untreated arm receives more bites than if offered alone. This is the concept of diversion but it has never been documented with mosquitoes in the field.

In the field, the notion of diversion is being tested using a mark-release-recapture design of *Ae. aegypti* within a cluster of experimental huts containing both treated and untreated structures similar to the diagram in Figure 5. Marked cohorts of mosquitoes of a specific color are released outdoors at each hut and they are recaptured in entrance interception traps. By applying a spatial repellent chemical to one or several of the huts while leaving others untreated, one would assume that larger numbers of mosquitoes released at the spatial repellent treated hut would be found at the untreated huts (i.e. they are diverted or driven from the treated hut to an untreated source). However, this has not been the case to date. Baseline data suggests that when the huts are all left untreated, there is a low level of mixing in the recaptured populations at all huts. When a repellent is applied to one or several huts, the number of mosquitoes entering these huts declines but there is no significant increase in the numbers of mosquitoes from these huts that seek out the untreated hut as compared to baseline. While this data suggests that no diversion occurs with *Ae. aegypti* under these experimental conditions, additional work is required with other species and in other experimental situations.

Future of Spatial Repellents

The future of behavior-modifying chemicals and particularly spatial repellents for vector control relies on researchers to prove their importance in stopping pathogen transmission. Currently, the information we have for spatial repellency relies heavily on the historical data from DDT use. While this data is very compelling, DDT remains a chemical shrouded in controversy and anything associated with it is often discounted. The fact remains that it is still the most successful vector control chemical ever used, and it continues to function in areas where DDT-resistance is high. This is primarily due to its strong spatial repellent properties. The current charge to the scientific community is to 1) prove that behavioral actions can break human-vector contact and prevent pathogen transmission and 2) promote a new paradigm in vector control that will encourage industry to screen their chemical libraries for compounds that elicit these actions. Progress is being made but there are still many gaps in our understanding of the benefits and pitfalls of contact irritant and spatial repellent actions. With so much emphasis being placed on finding new tools and strategies that will reduce the global burden from vector-borne diseases, it would be folly to overlook this novel technique that has its foundation in historical data with DDT. The methodologies for increasing our chance of discovery are advancing by the day, in addition to the fact that laboratory and field protocols are being developed to standardize the evaluation of new products and platforms of delivery. The time is right to look at spatial repellency as a viable tool for disrupting pathogen transmission and expand our options for chemical control beyond the traditional toxicity paradigm.

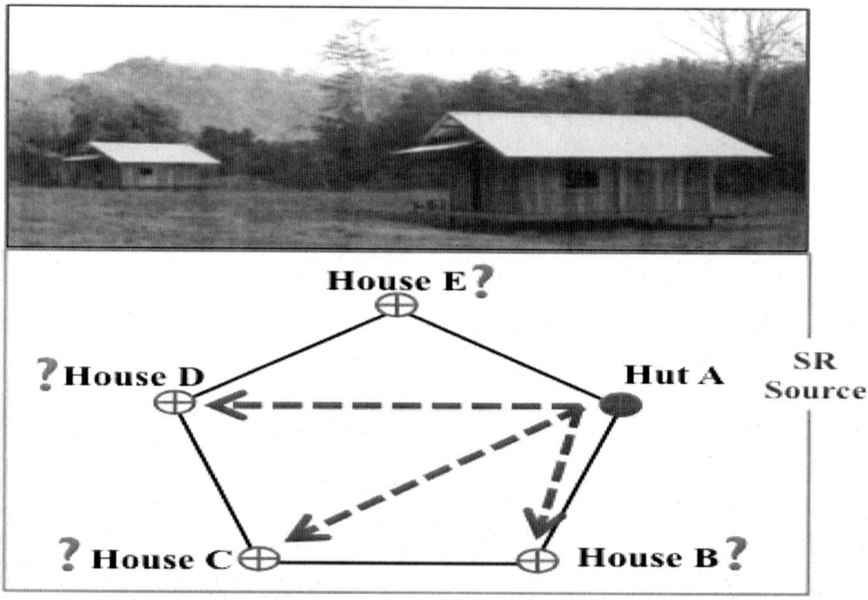

Figure 5. Schematic depicting the study design for the evaluation of diversion in an experimental hut study using Ae. aegypti in Thailand. In this setup, Hut A is treated with a spatial repellent, and a population of Ae. aegypti marked with a unique florescent marking powder is released at that hut and the mosquito movement is monitored to determine if marked mosquitoes are recaptured at the untreated huts.

References

1. Grieco, J. P.; Achee, N. L.; Sardelis, M. R.; Chauhan, R. K.; Roberts, D. R. A novel high-throughput screening system to evaluate the behavioral response of adult mosquitoes to chemicals. *J. Am. Mosq. Control Assoc.* **2005**, *21*, 404–411.
2. Achee, N. L.; Sardelis, M. R.; Dusfour, I.; Chauhan, K. R.; Grieco, J. P. Characterization of spatial repellent, contact irritant, and toxicant chemical actions of standard vector control compounds. *J. Am. Mosq. Control Assoc.* **2009**, *25*, 156–167.

3. Hill, N.; Lenglet, A.; Arnez, A. M.; Carneiro, I. Plant based insect repellent and insecticide treated bed nets to protect against malaria in areas of early evening biting vectors: Double blind randomized placebo controlled clinical trial in the Bolivian Amazon. *BMJ (Br. Med. J.)* **2007**, *335*, 1023–1026.
4. Kennedy, J. S. The excitant and repellent effects on mosquitoes of sublethal contacts with DDT. *Bull. Entomol. Res.* **1947**, *37*, 593–607.
5. Dethier, V. G.; Browne, L. B.; Smith, C. N. The designation of chemicals in terms of the responses they elicit for insects. *J. Econ. Ent.* **1960**, *53*, 134–136.
6. Muirhead-Thomson, R. C. In *Mosquito Behavior in Relation to Malaria Transmission and Control in the Tropics*; Edward Arnold and Company: London, 1953; p 219.
7. *Malaria Vector Control and Personal Protection*; WHO Technical Report Series 936; WHO (World Health Organization): Geneva, Switzerland, 2006.
8. Shreck, C. E.; McGovern, T. P. Repellents and other personal protection strategies against *Aedes albopictus*. *J. Am. Mosq. Control Assoc.* **1989**, *5*, 247–250.
9. *Testing Procedures for Insecticide Resistance Monitoring in Malaria Vectors, Bio-Efficacy and Persistence of Insecticides on Treated Surfaces*; Document WHO/CDS/CPC/MAL/98.12; WHO (World Health Organization): Geneva, Switzerland, 1998.
10. Brogdon, W. G.; McAllister, J. C. Simplification of adult mosquito bioassays through use of time-mortality determinations in glass bottles. *J. Am. Mosq. Control Assoc.* **1998**, *14*, 159–164.
11. Kramer, W. L.; Mulla, M. S. Oviposition attractants and repellents of mosquitoes: Oviposition response of *Culex* mosquitoes to organic infusions. *Environ. Entomol.* **1979**, *8*, 1111–1117.
12. Grieco, J. P.; Achee, N. L.; Chareonviriyaphap, T.; Suwonkerd, W.; Chauhan, K.; Sardelis, M. R.; Roberts, D. R. A new classification system for the actions of IRS chemicals traditionally used for malaria control. *PLoS ONE* **2007**, *8*, e716, http://www.plosone.org/article/info:doi%2F10.1371%2Fjournal.pone.0000716 (accessed August 8, 2007).
13. Roberts, D. R.; Alecrim, W. D.; Hshieh, P.; Grieco, J. P.; Bangs, M.; Andre, R. G.; Chareonviriyaphap, T. A probability model of vector behavior: Effects of DDT repellency, irritancy, and toxicity in malaria control. *J. Vector Ecol.* **2000**, *25*, 48–61.
14. WHO Recommended Insecticides for Indoor Residual Spraying Against Malaria Vectors. WHO (World Health Organization). http://apps.who.int/malaria/cmc_upload/0/000/012/604/IRSInsecticides.html (accessed September 2, 2009).
15. WHO Recommended Insecticide Products for Treatment of Mosquito Nets for Malaria Vector Control. WHO (World Health Organization). http://appswho.int/malaria/cmc_upload/0/000/012/605/ITNTable.htm.Journal (accessed September 2, 2009).
16. Yap, H. H.; Tan, H. T.; Yahaya, A. M.; Baba, R.; Loh, P. Y.; Chong, N. L. Field efficacy of mosquito coil formulations containing d-allethrin and

d-transallethrin against indoor mosquitos especially *Culex quinquefasciatus* Say. *Southeast Asian J. Trop. Med. Public Health* **1990**, *21*, 558–563.

17. Birley, M. H.; Mutero, C. M.; Turner, I. F.; Chadwick, P. R. The effectiveness of mosquito coils containing esbiothrin under laboratory and field conditions. *Ann. Trop. Med. Parasitol.* **1987**, *81*, 163–171.
18. Mosha, F. W.; Njau, R. J.; Alfred, J. Efficacy of esbiothrin mosquito coils at community level in northern Tanzania. *Med. Vet. Entomol.* **1992**, *6*, 44–46.
19. Amalraj, D. D.; Sivagnaname, N.; Boopathidoss, P. S.; Das, P. K. Bioefficacy of mosquito mat, coil and dispenser formulations containing allethrin group of synthetic pyrethroids against mosquito vectors. *J. Commun. Dis.* **1996**, *28*, 85–93.
20. Hemingway, J.; Beaty, B. J.; Rowland, M.; Scott, T. W.; Sharp, B. L. The innovative vector control consortium: Improved control of mosquito-borne diseases. *Trends Parasitol.* **2006**, *22*, 308–312.
21. Ditzen, M.; Pellegrino, M.; Vosshall, L. B. Insect odorant receptors are molecular targets of the insect repellent DEET. *Science* **2008**, *319*, 1838–1842.
22. Sharma, S. N.; Shukla, R. P.; Raghavendra, K.; Subbarao, S. K. Impact of DDT spraying on malaria transmission in Bareilly District, Uttar Pradesh, India. *J. Vector Borne Dis.* **2005**, *42*, 54–60.

Chapter 9

Repellents for Protection from Bed Bugs: The Need, the Candidates, Safety Challenges, Test Methods, and the Chance of Success

Robin G. Todd*

ICR, Inc., 1330 Dillon Heights Avenue, Baltimore, Maryland 21228-1199
***E-mail: RTodd@icrlab.com**

Owing to the lack of effective products for bed bug control, there is a need for repellents to protect people while they sleep. Repellents in general are discussed. It is concluded that only repellents applied to bedding fabrics and suitcases will be appropriate for protection from bed bugs. Candidate bed bug repellents and the challenges they face are summairzed. Repellents applied to fabric can be tested economically by *in vitro* methods, but more costly *in vivo* methods are needed for their greater realism. The following methods are described: petri dish assays, containers with fabric shelters, containers with attractants (heat or carbon dioxide), and a surrogate method using rabbits instead of humans. The use of human subjects, although ideal, for testing is unlikely.

Introduction

Until the 1950's bed bugs (*Cimex lectularius* L.), such as the one shown in Figure 1, were a common household pest in the US and many other parts of the developed world (*1*). Following the introduction of DDT and other highly effective chlorinated hydrocarbon insecticides, the incidence of bed bugs in this country fell to insignificant levels. So much so that, in the 1980's few entomologists had encountered this pest in the wild (*2*). Since the turn of the century, this picture has changed dramatically: there has been a huge resurgence of bed bugs in all parts of the USA and other parts of the world. Bed bugs are now the most commonly cited pest by pest management professionals and are amongst the most difficult to control; two or three visits are usually required by the PMP to achieve control (*2*).

Figure 1. An adult female bed bug. (photograph by Timothy Foard)

Causes of Resurgence of Bed Bugs in the US

Three likely reasons for the resurgence of bed bugs are given below.

i) Firstly there was the loss of DDT in 1972 and other chlorinated hydrocarbon insecticides. These chemicals were largely replaced by products containing organophosphates and carbamates. The use of such products has been almost entirely eliminated by USEPA's re-registration program, which began in 1988 (*3*) These products were replaced with pyrethroid-based products during the 1980's and 1990's. Bed bugs have since developed high levels of resistance to pyrethroids in many parts of the country (*4*).

ii) Ant and cockroach control was very often carried out by applying wet sprays from manually pressurized sprayers to base boards and other indoor surfaces. This approach caused considerable exposure to residents of treated buildings and this was therefore viewed by pesticide regulators as a safety problem. Such 'base board' spraying probably did, however, give unintended control of bed bugs (*2*). From the 1980's onwards control of these two pest groups has increasingly been by the use of toxic baits, either applied in the manner of caulking into cracks or in bait stations. Since bed bugs feed only on blood and since baits rely upon the insect ingesting the bait, this approach probably has no impact on *C. lectularius*.

iii) The increase in air travel has resulted in far more people arriving far more quickly in the US from countries where bed bugs had remained prevalent.

The Need for Repellents for Protection from Bed Bugs: Topical, Spatial, or Fabric Treatments?

As noted above, resistance to pyrethroids is one of the principal obstacles to bed bug control. While a few non-pyrethroid insecticides have been approved for bed bug control (examples are chlorfenapyr, a pyrrole, and hydroprene, an insect growth regulator), the bed bug problem shows no sign of abating.

Based on the increasing incidence of bed bugs, there appears to be a lack of effective, EPA-approved chemicals for killing bed bugs. There is therefore a need for alternative means of protecting people from this pest. Insect repellents, as defined by Dethier et al (*5*) work, not by killing the target insect, but causing them to 'make oriented movements away from' the repellents. Effective repellents would thus protect people from bed bugs by non-toxic means.

For protection against mosquitoes, biting flies, ticks and other haematophagous arthropods, several effective repellents are approved for application to the skin. Bed bugs typically bite at night, while their human hosts are asleep. Topically-applied repellents can be sticky (especially DEET-based ones) and are therefore not suitable for use prior to retiring for a night's sleep. In addition, such treatment would lead to prolonged and repeated human exposure; this would require the repellent to be extremely safe and would severely limit the pool of candidates.

Other types of repellents are the spatial repellent devices which emit vapor or smoke to keep insects, typically mosquitoes, out of the the immediate vicinity. Mosquito coils, mats, candles and emanators rely upon heat to mobilize the repellent. Typically pyrethrins or a pyrethroid (usually one of the allethrins) are the active ingredient. Although these are insecticides, they appear to act as repellents at sub-lethal concentrations (*6*), the use of spatial repellents against bed bugs has not been investigated. As with topical repellents, the use of such devices would result in prolonged and repeated exposure to people as they slept, raising safety concerns.

A third means of applying repellents to thwart bed bugs would be to treat fabrics or other substrates around the bed to protect people as they sleep. Thus the periphery of the mattress and box spring, other bedding items, the floor (whether carpeted or bare) beneath the bed could be treated. If effective, such treatments would allow the sleeper to be protected without the need for topical application and with commensurately less exposure. Other substrate applications of repellents could be to sofas and upholstered chairs. The treatment of night attire (pajamas, nightgowns etc) would increase the level of exposure to that of topical treatments and is probably best avoided. In addition, such treated clothing would not protect the wearer's feet, hands, face and other exposed areas.

Currently the only chemical approved for treatment of clothing to protect against arthropods such as mosquitoes and ticks is the pyrethroid, permethrin. This use was pioneered by the US military (*7*) for its uniforms. Permethrin is primarily an insecticide but is used in this context more as a repellent. Owing to the advent of resistance to pyrethroids in bed bugs, its value as a repellent for this pest is open to question.

Treatment of bedding or other fabrics could be by the consumer, using a hand-held aerosol for example, or impregnation at the time of manufacture. The latter approach would however require that the treatment survive washing of the bedding items.

Few compounds can survive laundering, although encasements for mattresses and box springs would require only occasional washing. Many candidate repellents are plant-derived and are not persistent. If a repellent gave only transient protection from bed bugs, as it would be needed to provide at least 8 hours of protection for night's sleep, application each night before retiring, seems feasible.

Suitcases could be treated with repellents to prevent travelers from carrying bed bugs and their eggs, which would otherwise create new infestations.

At the time of writing (March 2011), repellents applied to substrates appear to represent a new product category for bed bug protection. They would not control infestations since they would only be keeping these pests at bay, not directly killing them or otherwise affecting their numbers. But this is a worthwhile goal for those living in heavily infested buildings and for travelers faced with spending nights on the road in motels and hotels which could be infested.

Challenges

Products applied to fabric as bed bug repellents face several challenges if they are to be effective and of practical value. Bed bug repellents applied to fabric must be of low toxicity if they are to be approved by US EPA. This Agency typically requires a margin of exposure, commonly referred to as an MOE (formerly margin of safety or MOS) as shown below.

$$MOE = NOEL / Exposure$$

Where:

- NOEL = No Observable Effect Level (from laboratory study with the most sensitive animal species, via the most relevant route of exposure, in this case dermal exposure)
- Exposure = the maximum predicted level of exposure
- The units are mg active ingredient/kg of animal body weight/day. EPA normally requires an MOE of at least 100X, preferably 1000X.

For pre-impregnated fabrics, such as bedding items, repellents must be able to withstand washing. Permethrin-impregnated military uniforms designed to protect against mosquitoes , ticks and other pests, are able to withstand 25 or even 50 laundering cycles. Such repellents must be able to prevent hungry bed bugs from crawling towards a sleeping person. Hungry bed bugs will have a strong incentive to cross treated fabrics to reach what will often be their only food source.

Candidate Repellents

As noted previously, most repellents have been developed for topical application for protection against mosquitoes, biting flies, ticks and other biting arthropod pests groups. One of the most widely used and effective is DEET (N,N-diethyl m-toluamide); it was sythesized in 1953 and has been in use ever since (*8*). More recently developed repellents are picaridin (1-piperidinecarboxylic acid 2-(2-hydroxyethyl)-1-methylpropylester) and IR-3535 (3-[N-Butyl-N-acetyl]-aminopropionic acid, ethyl ester), p-menthane diol, citronella, geraniol and other plant-derived compounds.

Repellents in or on fabrics for protection against bed bugs must be persistent and, as noted above, resistant to washing. This is particularly true of fabrics which are impregnated by the manufacturer.

Naturally-occurring repellents can be highly volatile and are thus unlikely to endure for more than a few days. Such repellents would be best applied directly by the user before going to bed each night or at intervals of every few nights to replenish the repellent.

Permethrin acts as a repellent when applied to fabric, although its mode of action is insecticidal. Other conventional insecticides would probably also repel bed bugs, if the insects had the choice to avoid them, but these chemicals are not approved by EPA for uses involving considerable human exposure.

Methods of Testing Candidate Bed Bug Repellents

The first step in developing a repellent for use against bed bugs should be to test it for efficacy. Almost all candidate compounds tested will ultimately prove unsuitable as bed bug repellents, owing to poor efficacy, cost, safety or other reason(s). Therefore, to avoid wasting time and money, testing should start with the cheapest (but least realistic) methods. If repellency these results are promising, more realistic, but costly, test methods should be used.

The methods discussed are those which are used at ICR, Inc. Other contract and university laboratories may have other methods. This chapter does not include these, but the author wishes to make clear that other methods may be in use.

Bed bugs for repellent testing should be starved so that they will tend to search for a host. Three weeks without a blood meal has been used in many ICR tests.

In Vitro Tests

Candidate repellents should first be tested by *in vitro* methods since these are lest than *in vivo* ones. The following methods are in order of increasing complexity, realism and cost.

Arena Method – The First Step

This method is based on that used by Wigglesworth (*9*) for assessing the behavior of body lice, another crawling, blood-sucking ectoparasite of humans. The test is run in an unlit room with no air flow. Filter papers are cut into halves and one half is treated with the repellent and allowed to dry. Both halves are placed in a petri dish ('arena') so that their cut edges meet and they cover the floor of the dish. Five or ten such dishes are set up for each repellent variable to be tested. A separate group petri dishes is set up with one half treated with the diluent. Ten adult bed bugs are then placed on the untreated filter papers in each arena. Bed bug are recorded as on the untreated or on the treated filter paper halves at intervals of, typically 0.5, 2 and 24 hours (see Figure 2). The +24 hour reading allows an overnight period of darkness to when bed bug host seeking behavior is at its maximum.

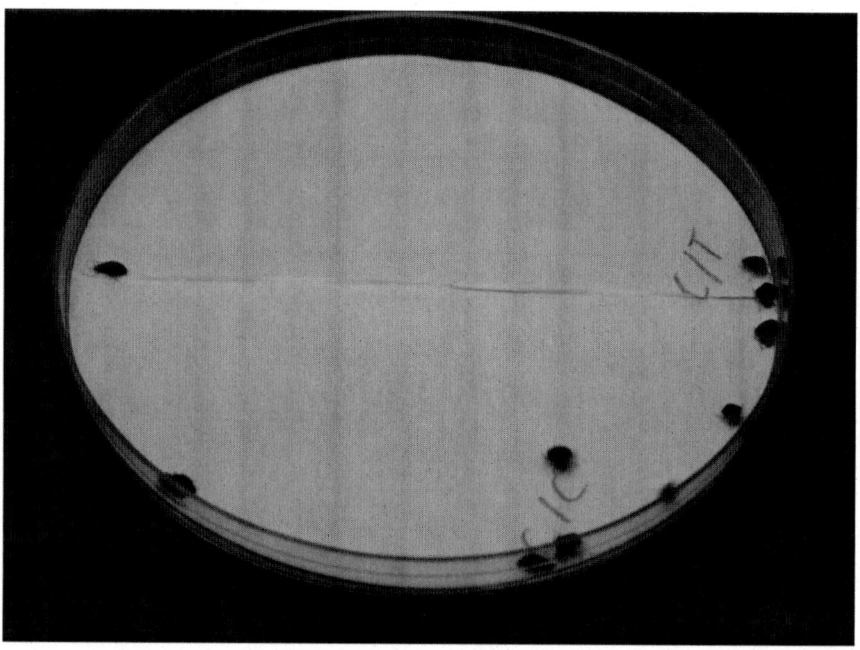

Figure 2. Arena in vitro method; bed bugs in place on a control arena (C/T is a filter paper section in same relative position as treated section in a test arena).

The numbers of bed bugs on the treated filter paper sections in the test arenas are compared with those on the positionally equivalent halves in the control arenas according to the Abbott's formula (*10*):

$$R = [\frac{C - T}{C}] * 100$$

Where:

- R= % Repellency for each test arena
- C = mean numbers of bed bugs on filter paper halves, in all control arenas, positionally equivalent to treated halves in test arenas
- T = number of bed bugs on treated half in one test arena
- The mean of R for each test arena is calculated.

Compounds showing promise in this test should be considered for further testing. The method relies on bed bugs' tendancy to distribute themselves on both halves of the filter paper. If most or all bed bugs in the control dishes remain on the filter paper half on which they were released, the results from the treatment dishes need to be discarded. Since this assay includes no positive attractant for bed bugs to crawl onto the treated halves of the filter papers, a positive result does not necessarily predict repellency under more realistic conditions.

Treated Fabric Shelters Method

This method takes advantage of the bed bug's predilection for seeking shelter in confined spaces. Sections of fabric (e.g. 2 x 4 inch sections of mattress ticking) are treated with the candidate repellent and allowed to dry. Other sections are left untreated or are treated with a diluent only to serve as the untreated controls. The sections are then folded in half and secured with a paper clip or similar to create shelters for bed bugs in the test. A treated shelter and an untreated shelter is placed on the floor of a container (see Figure 3). Pairs of untreated shelters are similarly placed in five other such containers. Groups of bed bugs (e.g. 10) are released on the floor of each container between the shelters. The containers are left for 24 hours. The numbers of bed bugs in or on each shelter are recorded, as well as any which are on the container floors. The latter category are regarded as non-participants in the test; the larger this group is, the weaker is the test result.The percentage repellency is calculated as in the preceding method.

Figure 3. In vitro repellent test set up with treated and untreated (control) shelters of mattress ticking.

In Vitro Test with Simulated Host Sensory Cues

Neither of the preceding assays employ host sensory cues. Incorporating such cues in a test will increase the challenge to potential repellents. Heat or carbon dioxide will achieve this without the expense and time involved in using live hosts. Both methods use shelters as described in the preceindg method.

Heat as an Attractant

With heat as the sensory stimulus, containers measuring ca. 18 x 24 inches are used as the test arenas. As above each is replicated (e.g. 5 containers). Heat is supplied by electric warming pads. The heating pad is placed under one end of the each container. Groups of 10 bed bugs are placed in shelters at the end of the container away from the heater pad. In addition sections of filter paper, obtained from a bed bug colony,are added. These papers carry the scent from many bed bug and are used as an added source of attractancy. Treated, empty shelters are placed at the end above the heater pads. An equal number of control containers are also set up; they are identical to the test containers, except that the shelters above the warming pads are untreated. The containers are left overnight. At +24 hours, the numbers of bed bugs found in the shelters are recorded. Bed bugs found in the treated shelters are evidence for repellent failure. Percent repellency is calculated as described above.

Carbon Dioxide as an Attractant

Many insects orient to carbon dioxide (CO_2) as an indicator of food resources. CO_2 is best known as an attractant of haematophagous arthropods; including bed bugs (*11*), (*12*), (*13*) and (*14*). It has been demonstrated fairly conclusively that CO_2 is the predominant attractant over heat and chemical attractants (*13*). Furthermore, they show that high CO_2 release rates (600-800 mL per min) calculated from amount of dry ice placed in thermos containers and left overnight, result in higher trap catch, while lower release rates (closer to 50% output of an adult human) are less likely to overwhelm the bed bugs within arenas and perform better in behavioral assays.

ICR has not yet developed a CO2-based test for bed bug repellents so the following account is conceptual.

When carbon dioxide is used as the sensory stimulant, a larger container is needed to allow the bed bugs to identify the concentration gradient of this gas so that they can orient towards the source. Children's splash pools are suitable as their smooth, steeply sloping sides are difficult for bed bugs to climb up, are cheap and are easily cleaned. It is advisable to create screened ports along the bottom side of the arena to allow some of the CO2 to escape and thus prevent prevent it from accumulating at unnaturally high concentrations. To accurately meter the CO_2, it is best to work with bottled gas (e.g. thr type used for paint ball guns), a regulator (set to 15-20 psi) , and a restrictor (0.006 inches) on the line supplying the traps. The output can be measured either with a flow meter or simply by inverting a graduated cylinder full of water into a bucket of water. The CO_2 line is then inserted into the cylinder so the gas bubbles to the top. By measuring the time it takes to fill a given volume of the cylinder with gas, one can then calculate the number of mL per minute. Carbon dioxide release rates should be around 280-325 mL per minute (30-35g/hour) to simulate the CO2 exhaled by a sleeping person. Sections of treated fabric are placed on platforms raised above the floors of the pools. The edges of these sections hang down and touch the floor. Other pools have untreated fabric set on platforms to serve as controls. The ends of the tubing are placed beneath the fabric so that the carbon dioxide flows out from it and downwards on to the floor. Bed bugs are placed in shelters which are set upon the pool floors on the far side from the platforms. The carbon dioxide flow provides the bed bugs with a plume to follow to the fabric sections.

The test set up is left out over night. The following morning the numbers of bed bugs found on or in the fabric sections on the platforms are recorded. Repellency is calculated as above. The higher the percentage of insects found on the control fabric sections, the more reliable is the value of any repellency calculated.

In Vivo Tests

Surrogate Hosts – An Optimal Compromise?

The use of human hosts for testing bed bug repellents will be difficult, as discussed in the following section. The use of animal surrogates, such as rabbits, in repellent testing would be far less demanding in terms of time, effort and cost. At ICR New Zealand white rabbits are used as the blood source for its bed bug colonies. These rabbits therefore represent the normal hosts for the ICR bed bugs, which adds to the rigor of the method.

The protocol for the proposed use of rabbits in repellent testing of fabrics, must be reviewed and approved by an IACUC (Institutional Animal Care and Use Committee), as required under USDA/APHIS regulations. It is important to note, however, that this method does not expose the rabbits to bed bugs or the test products, as explained below.

The tests are conducted in childrens' splash pools (approximately 42 inches diameter floor x 8 inches deep) which contain the bed bugs and prevent escapes. The interior sloping surfaces of the pools are coated with Fluon® to further prevent bed bugs escaping. In addition a layer of double sided tape is placed around the upper rim of each pool as a further precaution. The floors of the splash pools are lined with white paper, taped down around the base of the sloping walls (Figure 4).

The rabbits are placed in plastic carrier cages (28 x 21 x 20 inches high); large enough for them to move about easily. Food (dry lab chow and two types of fresh vegetables) and water are provided. The cages have a grating above the floor to isolate the rabbits from their waste products. Each cage is mounted upon a low wooden stool with four 6 inch high legs.

A simulated bed (cardboard or wooden box) measuring approximately 30 x 18 inch x 12 inch high is placed in each splash pool. A layer of treated fabric is secured around the vertical sides of the boxes. A board (with four 6-inch high legs) is placed on top of the carton, with each leg in a pitfall trap (ClimbUp™ Insect Interceptor, from Susan McKnight, Inc.). A cage containing a rabbit is then placed on top of each board (Figure 4). Starved bed bugs are then released on the floor of each splash pool (for example 50 adults, starved for 14 days). A control arena is also be set up in the same manner, except that untreated fabric is secured around the sides of the box.

In order to reach a rabbit, bed bugs must crawl up the sides of the boxes and over the treated fabric. However, any bed bugs which do so (and are thus not repelled) will be captured in the pitfall traps before they can reach the rabbits above. The tests are set up at 5.00 pm and are left overnight to allow maximum bed bug activity, which occurs in the early hours of the morning. The following morning, the test arenas are searched for bed bugs and their locations are recorded. Figure 5 shows bed bugs on the morning after a test was initiated, distributed along the lower edge of the box as well as on top of it.

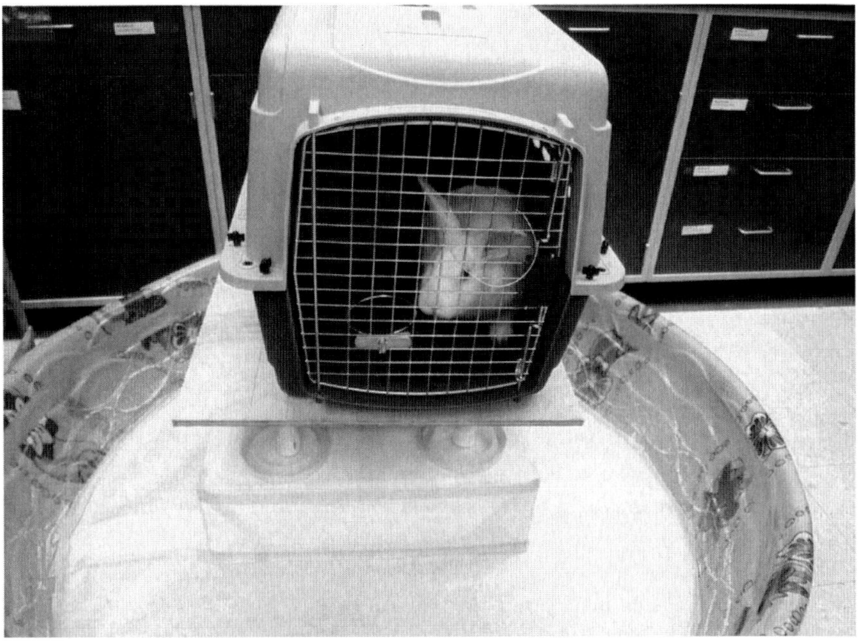

Figure 4. Rabbit in position (water bottle not present - it will be placed in loop).

Figure 5. Bed bugs at base and on top of surrogate bed (the latter indicate failure to repel).

Repellency is calculated by comparing the number of bed bugs in the pitfall traps, on the carpet tape or on the box in the (i.e. those which have not been repelled), with those found at the same locations in the control arenas. The greater the percentage of bed bugs in the control arenas which scale the sides of the surrogate bed, the stronger the resulting calculation of repellency (or lack of it).

Repellency is then calculated by the same formula as described previously for *in vitro* testing. The term'C' is the mean number of bed bugs in the four pitfall traps and on top of the box per pool (there maybe more than one), after one test night. The term 'T' is the number of bedbugs in the pitfall traps and on the top of the box in one test pool.

The test should be replicated for a total of at least three times, either on successive nights or with thrre pools per variable run concurrently. The rabbits should be switched such that control rabbits become test rabbits and the test rabbits become control rabbits so as to allow for any bias due to differences in their attractiveness to the insects. The identity and sex of the rabbits and the time since the bed bugs were last given a blood meal are noted in cases these influence the results.

Human Hosts – Ideal but Unlikely

The ideal test would be to test candidate bed bug repellents using human hosts as they slept at night. EPA's OPPTS 810.3700 Guidelines (*15*) outline requirements for testing repellents applied to human skin, but do not address treated fabrics. Testing with human subjects would be difficult and, as far as the author is aware, has never been done. Human subjects would need to be enrolled in a clinical trial. There is a strong possibility that EPA would require compliance with its 40 CFR part 26 Common Rule, which governs the use of human subjects in trials involving the deliberate exposure to pesticides. The Federal Insecticide, Fungicide & Rodenticide Act defines repellents as pesticides, despite their non-lethal mode of action. Repellents could be tested without exposing the subjects, so there is an argument for not invoking the arduous and length process needed to obtain EPA approval under the Common Rule.

To be meaningful, tests must be conducted overnight. Most potential test subjects will likely balk at the prospect of spending a night sleeping at a laboratory, on a bed mounted over an enclosure harboring hungry bed bugs, even if the subject cannot be bitten. There will be privacy concerns as well inconvenience. The author feels that human testing is an ideal which may not be realized.

Can Effective Repellents for Bed Bugs Be Developed?

Finally, it should be noted that, although an effective repellent for bed bugs would be a useful addition the existing set of product types, it may be that none will be found. If this transpires, it will likely be due to the innate drive of hungry bed bugs to seek a host and to the high threshold for safety of products for use on or around the bed. But it is important to find out. To date, preliminary data on naturally-derived, but proprietary compounds, tested at ICR by *in vitro* and the rabbit in vivo methods has revealed only modest repellency by the latter method.

Acknowledgments

I wish to thank the following ICR staff, past and present, for their invaluable support: Genoveva Collins, Reginald Coler, Timothy Foard, Ellen Quinn and Gloria Stephens.

References

1. Potter, M. F. The history of bed bug management – with lessons form the past. *Am. Entomol.* **2011**, *57* (1), 14–25.
2. Pinto, L. J.; Cooper, R.; Kraft, S. K. *Bed Bug Handbook. The Complete Guide to Bed Bugs and Their Control*; Pinto & Associates, Inc.: Mechanicsville, MD, 2007; p 1.
3. Pesticide Reregistration Facts, 2008. U.S. Environmental Protection Agency. http://www.epa.gov/oppsrrd1/reregistration/reregistration_facts.htm
4. Romero, A.; Potter, M. F.; Potter, D. A.; Haynes, K. F. Insecticide resistance in the bed bug: A factor in the pest's sudden resurgence? *J. Med. Entomol.* **2007**, *44* (2), 175–178.
5. Dethier, V. G.; Browne, L. B.; Smith, C. N. The designation of chemicals in terms of the responses they elicit from insects. *J. Econ. Entomol.* **1960**, *53* (1), 134–136.
6. Hudson, J. E.; Esozed, S. The effects of smoke from mosquito coils on *Anopheles gambiae* Giles and *Mansonia uniformis* (Theo.) in verandah-trap huts at Magugu, Tanzania. *Bull. Entomol. Res.* **1971**, *61*, 247–265.
7. McCain, W. W.; Leach, G. L. Repellents Used in Fabric: The Experience of the U.S. Military. In *Insect Repellents Principles. Methods, and Uses*; CRC Press: Boca Raton, FL, 2007; Chapter 13.
8. Moore, S. J.; Debboun, M. History of Insect Repellents. In *Insect Repellents Principles. Methods, and Uses*: CRC Press: Boca Raton, FL, 2007; Chapter 1.
9. Wigglesworth, V. B. The sensory physiology of the human louse *Pediculus humanus corporis* de Geer (Anoplura). *Parasitology* **1941**, *33*, 67–109.
10. Abbott, W. S. A method of computing the effectiveness of an insecticide. *J. Econ. Entomol.* **1925**, *18*, 265–7.
11. Rivnay, E. Studies in tropisms of the bed bug, *Cimex lectularius* L. *Parasitology* **1932**, *24*, 12–36.
12. Aboul-Nasr, A. E.; Erakey, M. A. S. On the behavior and sensory physiology of the bed-bug. I. Temperature reactions (Hemiptera: Cimicidae). *Bull. Soc. Entomol. Egypte* **1967**, *51*, 43–54.
13. Wang, C.; Gibb, T.; Bennett, G. W.; Mcknight, S. Bed bug (Heteroptera: Cimicidae) attraction to pitfall traps baited with carbon dioxide, heat, and chemical lure. *J. Econ. Entomol.* **2009**, *102* (4), 1580–1585.

14. Anderson, J. F.; Ferrandino, F. J.; Mcknight, S.; Nolen, J.; Miller, J. A carbon dioxide, heat and chemical lure trap for the bedbug, *Cimex lectularius*. *Med. Vet. Entomol.* **2009**, *23* (2), 99–105.
15. Product Performance Test Guidelines: Insect Repellents to be Applied to Human Skin; OPPTS 810-3700; U.S. Environmental Protection Agency: Washington, DC, July 7, 2010.

Chapter 10

Development of Essential Oil-Based Arthropod Repellent Products

Gretchen Paluch,*,[1] Steven Bessette,[1] and Roderick Bradbury[2]

[1]EcoSMART Technologies, Inc., 20 Mansell Court, Suite 375, Roswell, Georgia 30076
[2]EcoSafe Natural Products, Inc., #7-6782 Veyaness Road, Saanichton, British Columbia V8M 2C2, Canada
***E-mail: gpaluch@ecosmart.com**

The chemical ecology of plants provides a rich bank of structurally diverse compounds with a variety of insecticidal and repellent mechanisms. Further, the biological activity of select compounds offers potential for advances in pesticide chemistry with reduced risks to human health and the environment. Plant essential oil extracts represent the volatile essence of a plant, and are often comprised of a complex blend of terpenes and derivatives. Numerous studies have demonstrated that these compounds, as well as their parent blends, possess biological activity capable of eliciting adverse effects in arthropod pests. Recent patent literature has revealed novel research findings describing new arthropod repellent technologies and delivery applications; however, commercialization of essential oil-based products continues to lag behind. Factors affecting the commercialization of plant essential oil extracts as repellents, including regulatory requirements, intellectual property value, biological activity, product performance, and product quality, are discussed.

There is a growing demand for ecological products for managing arthropod pests (*1*) in today's society (*2–5*). Part of this demand is being met by the development of pesticide and repellent technologies that are classified by the EPA as reduced risk. Several consumer surveys report the demand for these types of products is driven by society's focus on health and safety, environment, and the psychological aspect of consumers' appearance consciousness, which has been shown to positively influence attitude toward personal care products (*6*). Data provided in the Harris Interactive survey, "Green Issues and Professional Pest Services" (*7*), on consumer sentiment to "green" products and services in the professional pest control industry further demonstrates that consumers are equally concerned with the safety of their family, as well as the environment. Consumers also provided perspective on a definition of "green" pest control services in their ranking of the top three most important characteristics: less toxic, biodegradable, and all-natural ingredients. Factors contributing to changes in the pesticide industry towards green pest control include the high costs associated with commercial licenses and dealing with unintentional exposures to conventional pesticides (*8*). As the market for botanical products continues to develop, the implementation of essential oil-based insecticides and repellents will depend greatly on guidelines that govern safe, green products and the evaluation of their efficacy and performance.

Regulation of Repellents Containing Essential Oils

In the United States, insecticides and repellents are regulated by the Environmental Protection Agency (EPA) in accordance with the Federal Insecticide Fungicide and Rodenticide Act (FIFRA) and the Federal Food Drug and Cosmetic Act (FFDCA). Pesticides are defined as "any substance or mixture of substances intended for preventing, destroying, repelling, or mitigating any pest, or intended for use as a plant growth regulator, defoliant, or desiccant" (*9*), and their registration is based on the conclusion that the pesticide will function without unreasonable adverse effects to humans or the environment (*10*). The risk to sensitive groups, especially findings pertaining to children in the 1993 National Academy of Science Report on Pesticides and Children, prompted the development of new standards contained in the 1996 Food Quality Protection Act (FQPA). This act changed the way pesticides and inerts were to be registered. Around the same time, there was renewed interest in biological pesticides. Current EPA classifications within this category include biochemical and microbial pesticides, microbial subspecies and strains of *Bacillus thuringiensis*, and plant-based pesticides. Reasons provided in the literature for this shift include reduced risks to human health and the environment associated with pesticide use, reduced regulatory burden for certain materials, and a delay or reduction of physiological resistance to conventional chemicals (*11*, *12*). In 1994, prior to the passage of FQPA, the EPA attempted to streamline the biological pesticide review process and advocate safe pest control. The Biopesticide and Pollution Prevention Division (BPPD) was created. In addition, the EPA's Pesticide Environmental Stewardship Program (PESP) supports the use of biopesticides as a key to integrated pest management (IPM) practices (*11*).

Classification as a biochemical pesticide requires the material to be derived from or be structurally similar and function identically to a natural substance, have a history of exposure to humans and the environment with minimal toxicity, and the substance must act by a nontoxic mode of action (*13*). As a result of the unique properties exhibited by biochemical pesticides, there are reduced data requirements for registration. Since the onset of the BPPD, a small number of plant-derived compounds and essential oil extracts have been registered as mosquito and tick repellents. These include 2-undecanone, dihydronepetalactone, citronella oil, lemon eucalyptus oil, and a component of lemon eucalyptus oil, p-methane-3,8-diol (Figure 1). In addition to EPA registration, research results on these materials have also been published (*14–17*).

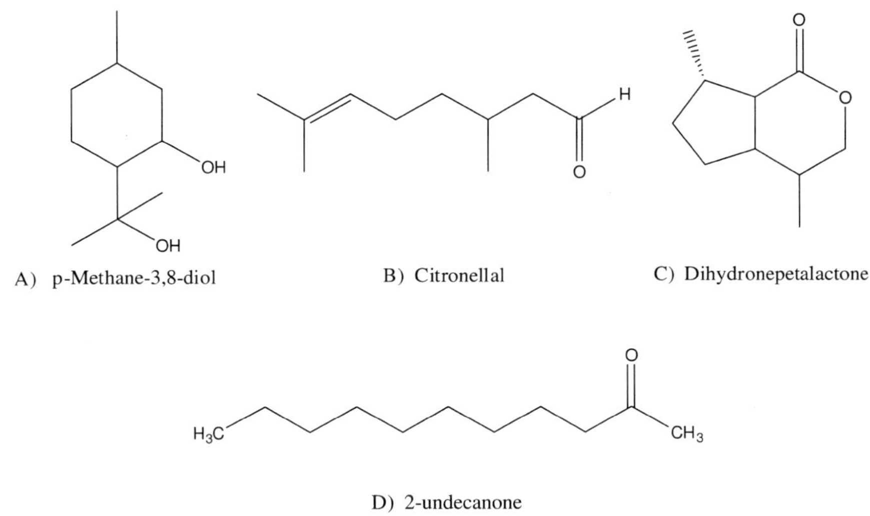

Figure 1. Biopesticide compounds registered as topical mosquito repellents. A) p-menthane-3,8-diol, from lemon eucalyptus oil, B) citronellal, from citronella oil, C) dihydronepetalactone, D) 2-undecanone.

Other plant essential oil extracts qualify as "minimum risk" pesticides and are exempt from registration as a result of EPA action under FIFRA, section 25(b) (*18*), on the basis that it is unnecessary to regulate the named substances in order to prevent unreasonable adverse human health and environmental effects for the intended use. Several factors were used by the EPA to assess the risk to human health and the environment for the intended use as a pesticide, including substances that are: 1) available to the general public (widespread) and have not shown evidence of adverse effects, 2) common food items, 3) FDA GRAS substances, 4) active via nontoxic modes of action, 5) no known data showing significant adverse effects to humans or the environment, 6) use patterns with negligible potential for incremental exposure, and 7) presumed to not persist in the environment (*19*). Current actives listed as exempt from EPA pesticide registration are provided in Table 1.

Table 1.
Active Ingredients Exempt from EPA Registration [FIFRA 25(b)] (*20*)

Castor oil	Linseed oil
Cedar oil	Malic acid
Cinnamon and cinnamon oil	Mint and mint oil
Citric acid	Peppermint and peppermint oil
Citronella and Citronella oil	2-Phenethyl propionate
Cloves and clove oil	Potassium sorbate
Corn gluten meal	Putrescent whole egg solids
Corn oil	Rosemary and rosemary oil
Cottonseed oil	Sesame and sesame oil
Dried Blood	Sodium chloride
Eugenol	Sodium lauryl sulfate
Garlic and garlic oil	Soybean oil
Geraniol	Thyme and thyme oil
Geranium oil	White pepper
Lauryl sulfate	Zinc metal strips
Lemongrass oil	

In order to meet the requirements for minimum-risk pesticides containing exempt active ingredients, products must also contain only EPA List 4A "Inert Ingredients of Minimal Concern" and meet requirements for maximum residue and tolerance limits for food use (*21*, *22*). Products must also abide by proper labeling requirements set forth by EPA. In the case of minimum risk pesticides for use as personal repellents, designed for protection against public health pests such as mosquitoes and ticks, there is movement toward additional data requirements to ensure effectiveness (*23*). Until such time, verification of product efficacy is limited to published findings in the scientific literature. Some companies have indicated that they have test data that supports their claims as efficacy data can be required for product registration at the state level.

Intellectual Property of Essential Oil Repellents

The literature on essential oil repellency in arthropods has been accompanied by similar growth in patent literature. Pohlit et al. reported in 2011 that since 1998, the number of patents for essential oil-based mosquito repellents almost doubled every four years (*24*). It is not surprising to see that most of the essential oil mosquito repellent patents are compositions, since there are subject matters that are excluded from patentability (*25*). Patentable subject matter can vary depending on legislation, but in some cases living organisms and organic substances may not be allowed. However, there are other categories of patentable materials that fall within the scope of essential oil repellents including but not limited to processes, material manipulated in process or ingredient, and apparatus/devices/delivery systems. Preceding these categories is the fact that inventions must be novel, useful, and exhibit some inventive step in order to qualify. Some of the more commonly listed essential oils in mosquito repellent patents are citronella (*Cymbopogon nardus*) and eucalyptus (*Eucalyptus* spp.) (*24*). Along with select natural compounds, additives, and compositions containing reference to combinations with synthetic repellents, insecticides or synergists (*24*), these represent some of the commonly pursued research, aimed toward commercialization.

Patents may be a source for chemical and biological properties, including biological activity of compounds or compositions, chemical synthesis methodology, properties of chemical additives (carriers, stabilizers), test methods, etc (*26*). Patents containing information on essential oil compositions potentially offer information on the synergistic effects of individual compounds since essentials oils are often mixtures of volatiles (example of complex mixtures from commercially available essential oils provided in Table 2).

The technology used to develop today's essential oil-based blends represent advances from their historic uses to control arthropod pests, as evidenced by the growing patent literature. Businesses that strive to develop such innovations look for protection and value of their inventions, in addition to meeting requirements for existing and emerging market opportunities. Other factors associated with commercialization and investments in essential oil-based technologies are based on the cost of product registration as well as the cost of production (*27*, *28*).

Biological Activity of Essential Oil-Based Repellents

Essential oil technologies are often associated with historical references to early plant extracts and compounds. For example, the utility of citronella oil for insect repellency was noted early in the 1900s (*30*) and is still a widely used plant extract for the protection against mosquitoes and other biting arthropods.

More current research on the range of essential oil-based repellents has been summarized in recent reviews (*31*, *32*). While the number of studies with structurally-related compounds and composition blends provide a basis for essential oil-based repellents, they do not always address the potency of individual essential oils and synergistic blends, nor do they provide guidance on how formulated products might work under field conditions.

Table 2. Common primary constituents of commercially available rosemary (*Rosmarinus officinalis)*, cinnamon (*Cinnamomum zeylanicum*), and peppermint (*Mentha piperita*) oils (%v/v) averaged from 30 samples (*29*)

Constituent	*Rosemary Oil*	*Peppermint Oil*	*Cinnamon Oil*
α-Pinene	10.8	-	-
Camphene	5.0	-	-
β-Pinene	6.2	-	-
3-Carene	0.4		
p-Cymene	1.3	-	-
d-Limonene	4.1	2.1	0.5
Linalool	1.2	-	
Linalyl acetate	14.6	-	
1,8-Cineole	37.8	5.1	-
Borneol	1.6	-	
Camphor	7.9	-	-
α-Terpineol	1.9	-	-
Caryophyllene	3.0	1.0	1.7
Bornyl acetate	1.5	-	
Cinnamaldehyde	-	1.3	2.2
Pulegone	-	1.9	
Eugenol	-	1.5	88
Menthone	-	25.7	-
Menthofuran	-	4.7	-
Menthyl acetate	-	5.2	
Menthol	-	48.8	-
Humulene			0.6
Benzyl benzonate			3.8
Minor compounds	2.7	2.7	3.2

Laboratory Trials To Screen EPA Exempt Minimum Risk Actives for Repellency

The behavioral response of the yellow fever mosquito (*Aedes aegypti*) to EPA exempt minimum risk actives was evaluated in a static-air glass apparatus as previously reported (*33*, *34*). Adult female mosquitoes were sourced from a laboratory colony in the Iowa State University Medical Entomology Laboratory, maintained at 80% relative humidity, held at 27°C, and fed a 10% (0.3 M) sucrose solution. The test apparatus consists of a 9 x 60-cm section of glass tubing with a 2-cm hole drilled at the midpoint along the length for central introduction of the insects. The position of the treated end, to the right or to the left, was selected by using a random-number table. Fifteen adult female mosquitoes were immobilized with CO_2 and then introduced to the 9 x 60 cm glass cylinder through the center 2-cm hole. All observations were completed in a temperature controlled room held at 26°C. Each active ingredient material was evaluated in the test chamber at a 0.5% (wt/wt) concentration in acetone solvent applied directly with a pipette to a filter paper with a surface area = 63.6 cm^2. Solvent was allowed 2 minutes to evaporate prior to placing the treated filter paper inside the test chamber. The rate of active ingredient on the treated filter paper surfaces was 78 $\mu g/cm^2$. Timing of behavioral responses began 1-2 minutes after mosquito introduction, once all mosquitoes recovered from the CO_2 immobilization, and ended at 15 minutes. Two measurements of repellency were made: % Repellency (spatial) and Contact repellency.

% Repellency= ((Number on Untreated-Number on Treated)/Total) X 100

% Repellency values were compared to the control using LS Means. Negative % repellency values would be representative of mosquito attraction to treated filter papers. Contact repellency was defined as 100% avoidance of the treated surface after 15 minutes (no contact = 100% avoidance). Contact repellency of the treatments were compared to the control using Fisher's Exact test.

The data presented in Table 3 represent initial or short-term mosquito spatial and contact repellency when exposed to surfaces treated with essential oils. Cinnamon, clove, eugenol, peppermint oil, rosemary oil, white thyme oil, and citronellal (individual compound from citronella oil served as natural standard for comparison) significantly repelled the mosquitoes under laboratory conditions. Out of the essential oil samples tested, peppermint oil showed the highest level of contact and % repellency (spatial) in the laboratory. In addition to the initial laboratory trials such as these, other methods are often employed to measure a reduction in biting rates (*35*, *36*). Results from field testing is less common, but there are studies published on protection from biting mosquitoes with lemon eucalyptus (*37*), palmarosa (*38*), and lemongrass oil (*39*). More recently, a number of studies have been conducted on the potential utility of area-wide tick suppression using isolated terpenes and/or parent essential oil extracts. Similarly, studies with lemon eucalyptus, geranium, lavender (*40*), rosemary and peppermint oils, nootkatone, carvacrol (*41*, *42*) and Alaskan yellow cedar (*43*) provide a baseline of activity for future improvements.

Table 3. Adult female yellow fever mosquitoes, *Aedes aegypti*, repellency to EPA exempt minimum risk actives measured after 15 minutes of exposure in a static-air repellency chamber to (*44*)

Treatment	*Average % Repellency*	*Contact Repellency*** *P-value*
Cedar Oil	33.3	0.2
Cinnamon Oil	82.2*	0.2
Citronella Java	19.7	0.2
Clove Oil	55.6*	0.2
Eugenol	82.2*	0.2
Lemongrass Oil	14.3	0.2
Peppermint Oil	95.5*	0.05
Rosemary Oil	60.0*	0.2
White Thyme Oil	51.1*	0.5
Natural Standard – Citronellal	77.8*	0.2
Control	-11.1	-

* Significantly differed from control (α=0.10). ** Mosquito contact on treated filter paper was recorded after 15 min and measured in 100% mosquito avoidance (no contact) from the treated surface. P-values are from Fisher's Exact test comparing each treatment to the control.

Assessing Performance and Quality of Essential-Oil Based Products

Essential oils are often comprised of a chemical blend of terpenoids that represents the volatile essence of a plant. Due to differences in blend complexity and processing methods such as isolation, extraction steps, and storage, the quality of essential oils from single species can vary greatly. Knowledge of the properties of individual components is needed to set standards for quality, performance, use profile (i.e. how the components will be used), product safety, and supply of raw materials. Setting standards to this end can be a challenging task, and begins with characterization of common arthropod responses to the individual compounds.

Select components may function as true repellents, which will cause the pests to orient or move away from a source material (*45*). This is only one behavioral effect, as there are other potential roles for active ingredient components. Today, the term "repellent" is used broadly to describe a variety of arthropod responses, depending on target pest and product application. In the case of products designed to mitigate or protect against hematophagous arthropods, the term "repellent" is used for any material that results in reduced biting rates of hematophagous arthropods (*46*). Since arthropods can exhibit a variety of behaviors and progress through different events prior to blood-feeding (including host-seeking and surface contact/agitation), alternate functionality of a compound, or multiple compounds, may prove useful in reducing overall biting rates. When considering the functionality of the individual compounds contained in an essential oil blend, or desired formulation, key considerations may include reduction of attractant short-range volatiles emitted by the host, reduced volatilization of repellent or deterrent compounds, and improved stability or fixative effects, such as those noted with the addition of vanillin (*17*, *32*).

One can also draw parallels to volatiles that act to protect plants against herbivores, or aid in pollination. A variety of more common ecological roles of terpenes produced by plants are summarized in the literature (*47*). One specific example of how select volatile components interact includes a series of studies conducted on wild tobacco (*Nicotiana attenuate*) floral volatiles (*48*). Wild tobacco is commonly pollinated by hummingbird moths. Important floral compounds for attracting and repelling pollinators were investigated. The studies identified benzyl acetone (attractant) and nicotine (repellent to reduce florivory and nectar robbing) were important floral volatiles required to maximize wild tobacco reproduction. Knowledge of such chemical interactions and ecological roles play a major part in product development of repellents.

Identification of the essential oil active ingredient and the key properties inherent to its individual components offers a challenge to formulation development in the repellent product category. An important step in the process is to establish a baseline of desired effects as they relate to performance standards for commercial product repellents, and translate this into a blend quality or composition standard. Further, establishment of the blend composition provides added benefits for product development and assessing the supply of raw ingredient essential oils. In some cases the complexity of the blend composition of essential oil actives requires sourcing high quality materials that is cost-prohibitive. However, high costs might be mitigated by improving such issues as plant cultivation conditions, production and post-production practices, and shipping.

Future Outlooks

The complexity of plant essential oil blends creates a unique challenge to the development of commercially acceptable arthropod repellents. Variability in the oils due to source and processing procedures poses a challenge in the development of a formulation and in the manufacturing process. Nevertheless, one could argue that both the familiarity and inherent complexity of the blend composition and its properties has helped to maintain interest over time, providing a large amount of foundational research, particularly with respect to use of the same essential oils in food and cosmetics.

The current trends in growth of essential oil-based repellent products highlight the demand and market for green products. The future outlook for these types of products is dependent on meeting industry guidelines that govern essential oil-based repellent products and protect returns on investments. At a minimum, these products should be labeled such that their safety and efficacy is made clear to consumers seeking pest management or personal protection products.

References

1. Moisander, J. *Int. J. Consum. Stud.* **2007**, *31*, 404–409.
2. Alder, J. *Newsweek* **2006**, July 17, 42–52.
3. Kangun, N.; Carlson, L.; Grove, S. J. *J. Public Policy Mark.* **1991**, *10*, 47–58.
4. Martin, B.; Simintiras, A. *Mark. Intell. Plann.* **1995**, *13*, 16–23.
5. Todd, A. M. *Ethics Environ.* **2004**, *9*, 86–102.
6. Kim, H. Y.; Chung, J. E. *J. Consum. Market.* **2011**, *28*, 40–47.
7. Harris Interactive. 'Green' Issues and Professional Pest Services. Prepared for the National Pest Management Association and presented at Going Green: Marketing to the 21st Centry Consumer, Denver, CO, 2007.
8. Tripathi, A. K.; Upadhyay, S.; Bhuiyan, M.; Bhattacharya, P. R. *J Pharm. Phytother.* **2009**, *1*, 52–63.
9. Code of Federal Regulations. Title 40, Part 152.15, July 1, 2010.
10. Federal Insecticide, Fungicide, and Rodenticide Act. 1996, 7 U.S.C. §136 et seq.
11. Anderson, J.; Leslie, A.; Matten, S.; Kumar, R. *Weed Technol.* **1996**, *10*, 966–968.
12. McClintock, J. T. Biopesticides Use and Delivery. In *Methods in Biotechnology*; Hall, F. R., Menn, J. J., Eds.; Humana Press, Inc.: Totowa, NJ, 1999; Vol. 5, pp 415–441.
13. Code of Federal Regulations. Title 40, Part 158.2000, July 1, 2010.
14. Witting-Bissinger, B. E.; Stumpf, C. F.; Donohue, K. V.; Apperson, C. S.; Roe, R. M. *J. Med. Entomol.* **2008**, *45*, 891–898.
15. Spero, N. C.; Gonzalez, Y. I.; Scialdone, M. A.; Hallahan, D. L. *J. Med. Entomol.* **2008**, *45*, 1080–1086.
16. Carroll, S. P.; Loye, J. *J. Am. Mosq. Control Assoc.* **2006**, *22*, 507–514.
17. Tawatsin, A.; Wratten, S. D.; Scott, R. R.; Thavara, U.; Techadamrongsin, Y. *J. Vector Ecol.* **2001**, *26*, 76–82.
18. Environmental Protection Agency. Exemption of Certain Pesticide Substances from Federal Insecticide, Fungicide, and Rodenticide Act Requirements. *Fed. Regist.* **1996**, *61*, 8876−8879.
19. Environmental Protection Agency. Pesticides; Exemption of Certain Substances from Federal Insecticide, Fungicide, and Rodenticide Act Requirements. *Fed. Regist.* **1994**, *59*, 49400−49401
20. Code of Federal Regulations. Title 40, Part 152.25(f), July 1, 2010.
21. Environmental Protection Agency. Inert Ingredients Eligible for FIFRA 25(b) Pesticide Products. http://www.epa.gov/opprd001/inerts/section25b_inerts.pdf (updated on December 20, 2010).
22. Code of Federal Regulations. Title 40, Part 180.1-2020, July 1, 2010.
23. Environmental Protection Agency. Petition to Amend FIFRA Section 25(b); Notice of Availability; Reopening Comment Period. *Fed. Regist.* **2006**, *71*, 70764.
24. Pohlit, A. M.; Lopes, N. P.; Gama, R. A.; Tadei, W. P.; de Andrade Neto, V. F. *Planta Med.* **2011**, *77*, 598–617.
25. World Trade Organization. Trade-Related Aspects of Intellectual Property Rigths, 1991, Article 27.3(b).

26. Howerton, P. W. *J. Chem. Doc.* **1964**, *4*, 232–234.
27. Semmler, M.; Abdel-Ghaffar, F.; Al-Rasheid, K.; Mehlhorn, H. *Parasitol. Res.* **2009**, *105*, 1483–1487.
28. Isman, M. B.; Miresmailli, S.; Machial, C. *Phytochem. Rev.* **2011**, *10*, 197–204.
29. Bradbury, R. 2010, unpublished report submitted to EcoSMART Technologies, Inc.
30. Smith, J. B. *N. J. Agric. Exp. Stn., Cook Coll.* **1901**, *542*, 37–42.
31. Isman, M. B. *Annu. Rev. Entomol.* **2006**, *51*, 45–66.
32. Maia, M. F.; Moore, S. J. *Malaria J.* **2011**, *10* (Supplement 1), S11.
33. Schultz, G. E.; Peterson, C.; Coats, J. Natural Insect Repellents: Activity against Mosquitoes and Cockroaches. In *Natural Products for Pest Management*; Rimando, A. M., Duke, S. O., Eds.; ACS Symposium Series 927; American Chemical Society: Washington, DC, 2006; pp 168–181.
34. Paluch, G. E.; Grodnitzky, J.; Bartholomay, L.; Coats, J. *J. Agric. Food Chem.* **2009**, *57*, 7618–7625.
35. Klun, J. A.; Debboun, M. *J. Med. Entomol.* **2000**, *37*, 177–181.
36. Klun, J. A.; Kramer, M.; Debboun, M. A. *J. Am. Mosq. Control Assoc.* **2005**, *21*, 64–70.
37. Moore, S. J.; Lenglet, A.; Hill, N. *J. Am. Mosq. Control Assoc.* **2002**, *18*, 107–110.
38. Ansari, M. A.; Razdan, R. K. *Indian J. Malariol.* **1994**, *31*, 95–102.
39. Moore, S. J.; Hill, N.; Ruiz, C.; Cammeron M. *J. Med. Entomol.* **2007**, *44*, 624–630.
40. Jaenson, T. G.; Garboui, S.; Palsson, K. *J. Med. Entomol.* **2006**, *43*, 731–736.
41. Jordan, R. A.; Dolan, M. C.; Piesman, J.; Schulze, T. L. *J. Econ. Entomol.* **2011**, *104*, 659–664.
42. Rand, P. W.; Lacombe, E. H.; Elias, S. P.; Lubelczyk, C. B.; St. Amand, T.; Smith, R. P., Jr. *J. Med. Entomol.* **2010**, *47*, 695–698.
43. Dolan, M. C.; Jordan, R. A.; Schulze, T. L.; Schulze, C. J.; Manning, M. C.; Ruffolo, D.; Schmidt, J. P.; Piesman, J.; Karchesy, J. J. *J. Econ. Entmol.* **2009**, *102*, 2316–2324.
44. Coats, J. R.; Schultz, G. 2004, unpublished report submitted to EcoSMART Technologies, Inc.
45. Dethier, V. G.; Browne; Barton, L. *J. Econ. Entomol.* **1960**, *53*, 134–136.
46. White, G. In *Insect Repellents Principles, Methods, and Uses*; Debboun, M., Frances, S. P., Strickman, D.;CRC Press: Boca Raton, FL, 2007; pp 31−46.
47. Gershenzon, J.; Dudareva, N. *Nat. Chem. Biol.* **2007**, *3*, 408.
48. Kessler, D.; Gase, K.; Baldwin, I. T. *Science* **2008**, *321*, 1200–1202.

Chapter 11

The Public Entomology Landscape: Development of Chemical Products against Biting Pests

Daniel Strickman*

Agricultural Research Service, Office of National Programs, U.S. Department of Agriculture, 5601 Sunnyside Ave., Beltsville, Maryland 20782
*E-mail: daniel.strickman@ars.usda.gov

The public generally have one desire when it comes to biting pests like mosquitoes and ticks: They don't want to be bitten. From an individual's standpoint, this sounds simple but from a technical standpoint the best result may require a complicated series of activities. The chemical industry has produced many products that help people avoid bites and it has often introduced them in ways that make the use of the product as straightforward, safe, and effective as possible. The best products have particular and specific roles in the process of controlling the pests. New development is likely to succeed when its designs fit logically into the larger process of pest control, often referred to as "integrated pest management" or IPM [(ref 1) Luckmann, W. H.; Metcalf, R. L. In Introduction to Insect Pest Management, 3rd ed.; Metcalf, R. L., Luckmann, W. H., Eds.; *Environmental Science and Technology*; John Wiley & Sons, Inc.: New York, 1994, pp 1−34].

The steps necessary to stop a biting problem can be divided in many different ways. It is helpful to categorize these steps in four stages: 1) Risk assessment – what is the problem? 2) Surveillance – where are the pests? 3) Control – how can bites be stopped? and 4) Sustainment—how can the solution to the problem be maintained? Other definitions of IPM have stressed the de-emphasis of the use of pesticides, the involvement of the user community, and the role of prevention. Regardless of the exact definition, the steps described above have to be executed to accomplish a reasonably integrated and targeted plan to reduce the impact of biting pests.

An entomologist is naturally going to stress identification as the first step in the solution of any problem. Trained to the standards of identification to species and always held to species identification for publication, an entomologist is never going to be completely comfortable with a program that does not place an accurate name on the arthropod that is the object of IPM. This viewpoint is supported by the valid point that the practitioner can only learn more from experience by knowing what the experience concerned. Also, there are examples of how the neglect of specific identification caused disastrous results by directing valuable resources on the wrong insect.

For research, recording experience, and often for effectiveness, it is necessary to identify the species of pest; however, actual pest control operations are often based on a more general identification. This is especially true when the measures to be taken against the insect are practically the same for any of a number of related species. Repellents are a good example, as the same product is often recommended for such widely different pests as ticks and mosquitoes. Those who work directly with the public may find that individual citizens would rather hear a common name that fits into his or her experience than an unfamiliar scientific name. The individual who receives the information from an entomologist is naturally most interested in solving the problem and being able to describe the problem as a story that can be related to others. As practitioners, those who give entomological advice to the public need to shape that advice around the expectations and capabilities of the individuals they serve in order to get the most effective result.

Recommendations start by assessing the problem, sometimes identifying the pest by descriptions over the telephone. At this level, the common names in Table 1 may be helpful objectives of the conversation. Each of those pests suggests a certain level of concern and therefore a certain level of what is reasonable to solve the problem. What is more, each of the common names in Table 1 denotes a group of arthropods the control of which is associated with a suite of effective products and strategies. This reasonable level of identification is not the only important information for product selection. In addition to knowing what the pest is, it is also necessary to know the situation. For example, a mosquito problem in a home located in a region with West Nile virus transmission and occupied by residents over 50 years old, who are most susceptible to serious consequences from infection, is more serious than a mosquito problem in a home where pathogen transmission is unlikely and the residents are younger. Table 2 summarizes pathogens transmitted by arthropods. The actual risk from disease is a combination of the likelihood of transmission in a particular location and the severity of the consequences. Malaria of any kind is a serious matter, but *falciparum* malaria frequently causes death in the non-immune. One could argue that it is strongly advisable to take precautions against mosquito bites where milder forms of malaria prevail, but that it is irresponsible not to take precautions where *falciparum* malaria is common. Scrub typhus, the rickettsial pathogen of which is transmitted by chiggers in Asia, is easily treated with readily-available antibiotics and transmission is associated with agricultural, rural, or disturbed habitats that are infrequently visited by tourists. The casual traveler to Southeast Asia does not need to worry about this disease unless he or she goes hiking or participates in rural life.

Table 1. Useful common names familiar to the English-speaking public and likely to result in accurate identification

Scorpion	Centipede	Sand fly	Green head
Spider	Louse	Mosquito	Stable fly
Mite	Bed bug	Snipe fly	Wasp
Chigger	Kissing bug	Horse fly	Ant
Tick	Biting midge	Deer fly	Bee

Table 2. Summary of arthropod-transmitted pathogens of humans. Some of the pathogens are grouped into similar forms; easily recognized common names are used for vectors (*2*).

Disease	*Kind of pathogen*	*Kind of arthropod vector*
Oropouche fever	Virus	Biting midge
Onchocerciasis, river blindness, craw craw	Round worm	Black fly
Endemic typhus, Plague	Bacteria	Flea
Loiasis: Calabar swelling, eye worm	Round worm	Horse fly
Chagas disease	Protozoa	Kissing bug
Epidemic typhus, Relapsing fever	Bacteria	Louse
Rickettsial pox, Scrub typhus	Bacteria	Mite

Continued on next page.

Table 2. (Continued). Summary of arthropod-transmitted pathogens of humans. Some of the pathogens are grouped into similar forms; easily recognized common names are used for vectors (*2*).

Disease	*Kind of pathogen*	*Kind of arthropod vector*
Bunyamwera, Bwamba, Ilesha, Pongola, California encephalitides, Jamestown Canyon, La Crosse, Snowshoe Hare, Tahyna, Chikungunya, o'nyong-nyong, Ross River, Dengue, Eastern equine encephalitis, Epidemic polyarthritis, Japanese encephalitis, Murray Valley, Australian X, Rift Valley Fever, Sindbis, Ockelbo, Pogosta disease, Karelian fever, St. Louis encephalitis, Venezuelan equine encephalitis, Western equine encephalitis, West Nile	Viruses	Mosquito
Filariasis, elephantiasis	Round worm	
Malaria	Protozoa	
Carrion's disease, oroya fever, verruga	Bacteria	Phlebotomine sand fly
Leishmaniasis: Kala azar, espundia, Baghdad boil	Protozoa	
Sand-fly fever, Phlebotomus fever	Virus	
Babesiosis	Protozoa	Tick
Colorado tick fever, Crimean-Congo hemmorhagic fever	Virus	
Ehrlichiosis, Lyme, Relapsing fever, Rocky Mountain spotted fever, Boutonneuse fever	Bacteria	

Knowing the problem and the relative risk it presents is, of course, just the beginning. Often the next step does not involve use of chemical products. If the source of the arthropod is on an individual's property, there is a good chance that the owner can limit or eliminate the source. Establishing good drainage, properly managing rainwater barrels and other intentionally stored water, and discarding abandoned containers are the usual measures recommended to eliminate sources of larval mosquitoes (*3*). Sealing holes in a structure can exclude rodents, bats, and birds from a house, where they might otherwise create problems with flies or mites. Poorly managed compost heaps, animal feces, and waste feed can become the sources of flies.

Chemical products can be a big help in eliminating sources of pests, especially when combined with other efforts to limit the sources of pests. Home owners have access to some products that are effective against mosquito larvae, including *Bacillus thuringiensis israelensis* (BTI) and methoprene. BTI (*4*) produces a complex mixture of protein toxins that disrupt the gut of mosquitoes and related flies, without any harm to vertebrates or even to unrelated insects. It is commonly formulated for the public in long-lasting granules or doughnut-shaped cakes that provide protection over a period of weeks. Methoprene (*5*) is a chemical that mimics the juvenile hormone of insects, resulting in failure of the larvae to complete development. Trials have shown remarkable specificity used at label dosages against mosquito larvae, possibly because the larvae develop more rapidly than many other kinds of aquatic insects. Methoprene is also sold to the public in a long-lasting formulation. Both of these products are safe and easy to use, but each has the disadvantages of only affecting the larvae during about the first two-thirds of their development and not affecting pupae at all. Price is difficult to evaluate. On the one hand, a small container of either product will treat a large area of water, often more than the consumer needs. On the other hand, units as packaged often cost somewhere between $10.00 and $30.00, which may be expensive enough to discourage some customers. There are other products for mosquito larval control, but the others available to the public probably need more documentation of effectiveness.

Rodenticides might also be considered as part of source control, especially if sanitation and structural changes have made a property less suitable for their shelter and development. It is quite possible at that point to kill every rat or mouse on the property with a rodenticide and then be free of the rodents and the pests they support for a long time. Many pest control practitioners would object to even this use of rodenticides, advocating instead the careful use of traps. Unlike the mosquito larval control products described above, many rodenticide active ingredients are highly toxic to all vertebrates. Concerns over secondary poisoning of birds of prey, dogs, or cats are real, not to mention the hazard to children. All the same, rodenticides can be used safely and effectively by the public. Most products are anti-coagulants, but there are other modes of action including nervous disruption, vitamin overdose, respiratory inhibition, and disruption of digestion. Table 3 lists some recent active ingredients, though this list is likely to change with new innovations and new regulations. The anti-coagulants have been favored, in part because they can be formulated at a very low dose that reduces the chance of accidental poisoning of pets and children. Any visit to an urban grocery or a rural feed store will show that the demand for household rodenticides remains strong – in fact, much stronger than for home-use mosquito larvicides.

Elimination of sources of pests is often inadequate for the home. Despite the best efforts of the individual, deer drop ticks on the property, mosquitoes fly in from the neighbors or public areas, etc. The next line of defense is to erect barriers against these pests. Those barriers may be physical, like screen doors, window screens, sound walls and roofs, or deer fences. The barriers can also be chemical. The most common barriers are directed at either deer ticks or flying insects. Typically, deer tick control (*7*) is accomplished by applying the long-lasting pyrethroid, permethrin, to a broad band around the edge of the yard. Applied effectively, this treatment can disrupt deer ticks with only one application per year. Residual insecticides can be applied to kill flying insects that rest on treated surfaces. The available active ingredients include an organophosphate (malathion), a carbamate (carbaryl), and a variety of pyrethroids (including permethrin, deltamethrin, bifenthrin, and lamda cyhalothrin). Of all the household applications of pesticides, these probably represent the greatest hazard to the individual and to the environment. If the consumer purchases a concentrate, he or she must take extra precautions during mixing. During application, the householder will be handling large volumes of material and must wear proper protective clothing, as well as clean up properly afterward. If applications are done poorly, there is a real chance of contamination, especially of surface water. Finally, like many outdoor applications of pesticides, there is the risk of killing non-target insects, including those that are beneficial pollinators and biological control agents. Considering these potential problems, application of outdoor residuals by the householders should be considered carefully against the magnitude of the problem. Prevention of a hypothetical pest may not be justified. If there is a real pest problem and application of an outdoor residual is desirable, then every effort should be made to follow label directions.

Barriers can fail. Mosquitoes get indoors despite screens, people bring bed bugs home, and untreated pets may support a flea or brown dog tick infestation. Products for use indoors come in at least eight different formulations, reviewed in Table 4. Most treatments for mosquitoes and flying insects are with aerosols sprayed directly in the air, which can be very useful to remove mosquitoes that have filtered into a home. It is less common in the United States to apply broad surface treatments with the intention of killing mosquitoes that rest on surfaces, though this is a standard method for controlling malaria vectors overseas. Flying insects can also be treated with DDVP (*8*), an organophosphate, impregnated into plastic strips. The limitation on this product is that it can only be used in rooms that are occupied less than four hours per day. More commonly, indoor treatments are directed at crawling insects. Among the biting pests in this category are the brown dog tick, rat mites, fowl mites, kissing bugs, and bed bugs. Insecticidal treatment (other than medicinal application to the individual) is usually not recommended for scabies mites and lice. The most commonly applied products are sprayed by the consumer along baseboards and in cracks and crevices created by wood work or in furniture. Application to clothing, areas where food is prepared, vehicles, and beds is possible, but the list of registered products is more limited (Table 5). Most of the registered products are pyrethroids, so that there is a real problem when a pest is pyrethroid resistant. Fortunately, there are alternative active ingredients for indoor use. Chlorfenapyr (*9*) (a pyrrole, that is activated in the insect by oxidation, disrupts oxidative phosphorylation in the mitochondria) and pyriproxyfen (*10*) (a powerful growth regulator that disrupts normal hormonal function in the insect) are recent additions.

Most people would probably agree that it is best to take action to eliminate the biting pests. Modifying the surroundings so that the pests cannot develop and killing the pests that develop anyway were discussed above. However, what if eliminating the pests is not possible or treatments do not work? The last line of defense is repellents. In one form or another, these have been used for thousands of years (*11*). The appeal is clear in that a product applied to the skin stops biting arthropods exactly where they would have done their damage. Unlike insecticides, they are carefully non-toxic and inherently designed to be used by individuals. Possibly because they are a consumer product, industry continues to have an interest in new active ingredients.

Table 3. Recent list of rodenticide active ingredients available to the public in the United States (*6*)

Active ingredient	*Mode of action*	*Advantage*	*Disadvantage*
Brodifacoum	2nd-generation anticoagulant	Single feed	Highly toxic
Bromadiolone	2nd-generation anticoagulant	Single feed	Highly toxic
Bromethalin	Neurotoxin	Single feed	Highly toxic
Cholecaliciferol	Vitamin D overdose	Safer for wildlife	Multiple feed
Chlorophacinone	1st –generation anticoagulant	Safer	Multiple feed
Diphacinone	1st –generation anticoagulant	Safer	Multiple feed
Warfarin	1st-generation anticoagulant	Safer	Multiple feed
Zinc phosphide	Phosphine gas stops respiration	Very effective	Highly toxic

The well-documented, commonly available active ingredients in the United States are DEET (*12*), picaridin (*13*), IR3535 (*14*), and *para*-menthane diol (PMD) (*15*). DEET was developed by the U.S. Department of Agriculture for the Department of Defense in the late 1940s. It was the result of a large screening program with the specific purpose of finding new repellent active ingredients to supplement or replace the best compounds of the time (ethyl hexanediol [Rutgers 612], indole, and dimethyl phthalate). DEET was introduced to the public in the 1950s and gradually came to dominate the market. It has only been in the last 15 years that major new active ingredients have been introduced. IR3535 was discovered based on an amino acid's structure, gaining the registration advantages of a biopesticide. It was much more popular in Europe before it was introduced into the United States. Picaridin was the result of a large developmental program by a private firm based on molecular modeling. PMD emerged from the study of a Chinese preparation from the lemon eucalyptus tree, again achieving the advantages of a biopesticide. PMD can be synthesized in pure form, though American products currently use botanical extracts. Each active ingredient has its own advantages and disadvantages (Table 6), but they perform similarly. Probably the main distinctions between products are their application characteristics, the appropriateness of packaging (e.g., aerosol, lotion, towlette, etc.) for the intended use, and duration of protection influenced by percentage of active ingredient.

Table 4. Formulation for indoor-use insecticides (*6*)

Formulation	*Description and use*	*Advantages and disadvantages*
Ready-to-use solution	Liquid pre-mixed at concentration for use in a sprayer and applied to crack and crevices or to a surface as a residual	Minimal handling, manufacturing control over dosage, bulky, more expensive
Concentrate	A liquid or powder mixed with water and applied from a sprayer	Usually cheaper, flexibility of range of dosages, less bulky, greater risk of exposure, possibility of gross overapplication
Pressurized spray	Aerosol can with pre-mixed formulation for application in the air or to surfaces	Minimal handling, simple to use, often formulated for maximum safety, bulky, expensive
Total release spray	Aerosol can with a valve that releases the entire contents	Minimal skill required for treatment, simple to use, controversy over safety and effectiveness
Dust	Powder mixed with insecticide for use in unexposed areas and electrical equipment	No handling of concentrate, often highly effective for its use, only safe way to treat electrical boxes, messy, dusting applicators not widely available
Solid phase vaporizer	Plastic impregnated with DDVP that slowly volatilizes throughout room, for use in minimally occupied rooms	Easy to use, effective against flying insects and crawling insects exposed for a long time, special care for safe use
Chalk or felt tip	Chalk or felt tip containing insecticide and meant to draw a treatment zone with visible indication	Easy to use and inherently accurate placement, difficult to get adequate amount of insecticide on target, potential for high exposure

The public appears to have a great appetite for botanically-based repellent active ingredients (*16*). Perception strongly supports the idea that botanical compounds are likely to be safer and environmentally friendlier than synthetics, even though scientific evidence for these conclusions may be lacking. Technically, there is a kind of elegance to exploring plants for repellent compounds because of the ecological importance of secondary plant chemicals for protection against herbivorous insects. Highly effective botanical active ingredients like PMD and 2-undecanone further justify exploration of plant extracts and chemicals known to be produced by plants. At least 36 plant species are known to produce compounds that provide short term repellency and at least another 41 species produce compounds that provide longer-lasting repellency. Unfortunately, at least 29 of these extracts are potentially hazardous, either as toxicants or skin sensitizing agents. The potential harm from a plant extract is sometimes disputed. For example, citronella is not registered in Canada or Germany based on its potential for causing skin problems; whereas, citronella is a common, though less effective, active ingredient in the United States. Unfortunately for both the consumer and the botanical industry, many products are available with poorly documented effectiveness and safety. This situation makes recommendation of botanical repellents difficult. Perhaps the safest advice to those who want a botanical repellent is to always use an EPA registered product and to be ready to switch products if a particular one proves to be ineffective.

Table 5. Insecticide active ingredients registered for indoor use under special circumstances (*6*)

Area	*Active ingredients*
Around food	Deltamethrin, d-*trans*-allethrin, esfenvalerate, phenothrin, prallethrin, pyrethrins
Bedding, mattresses	Deltamethrin, d-*trans*-allethrin, methoprene, permethrin, phenothrin, prallethrin, pyrethrins
Carpets	Detlamethrin, d-*trans*-allethrin, esfenvalerate, methoprene, permethrin, phenothrin, prallethrin, pyrethrins, pyriproxyfen, S-bioallethrin, tetramethrin, tralomethrin
Clothing	d-*trans*-allethrin, methoprene, permethrin, phenothrin, prallethrin, pyrethrins, tetramethrin
Pet areas	Bifenthrin, cyfluthrin, cyhalothrin, deltamethrin, d-*trans*-allethrin, esfenvalerate, methoprene, permethrin, phenothrin, prallethrin, pyrethrins, pyriproxyfen, tetramethrin, tralomethrin
Vehicles	Cyfluthrin, cypermethrin, S-bioallethrin, esfenvalerate, permethrin, phenothrin, prallethrin, pyrethrins, pyriproxyfen, S-bioallthrin, tetramethrin

Table 6. Summary of the advantages and disadvantages of the four major repellent active ingredients used in the United States (*6*)

Active ingredient	*Advantages*	*Disadvantages*
DEET	Cheap, long safety and evaluation history, broad spectrum protection	Oily, distinct odor, melts plastics, irritates eyes, not as effective against ticks, kissing bugs, malaria mosquitoes
Picaridin	Broad spectrum, does not melt plastics, low odor, not as oily, works at lower concentrations	More expensive, less experience with use, not as effective against ticks, some malaria mosquitoes, biting midges
IR3535	Extremely safe, long evaluation record, low odor, not oily, does not melt plastics, broad spectrum	Repellency sometimes fails at low concentrations
PMD	Good against malaria mosquitoes and ticks, botanical derivative	Only partially evaluated, some preparations have strong odor, irritates the eye

There are a number of other products designed to prevent bites that are difficult to fit into an IPM program. Traps attempt to remove mosquitoes and other flying, biting insects from a limited area. Area repellent products (*17*) attempt to disperse a chemical into the air to provide a bubble of protection around a person or a small group of people. Both kinds of products have the appeal of easy use and the perception of low exposure to chemicals. That perception is based on the lack of any residual application, personal distance from the device, and, sometimes, simply the familiarity of long use. Considerable developmental work has gone into the design of traps for mosquitoes, including work on physical and chemical attractants, the geometry of air flow, and visual attraction to mosquitoes. There is scope for improvement in all of these areas, especially chemical attractants, as none of the traps achieve the public's desired result of a turn-key solution to mosquito problems. Certainly, it would be a great success to produce a device that consistently attracted mosquitoes more effectively than a person over a large enough area to provide protection in a garden or patio. The public's desire for such a device is demonstrated by the large market for traps that are, at best, only partially effective.

Area repellents are another large market in mosquito protection. Wrist bands containing DEET or geraniol, electronic repellers, and ultrasonic emitters are inefficient at best and completely ineffective at worst, yet the public continue to buy them in quantity. Tragically, people continue to poison themselves by wearing dog or cat flea collars in an attempt to avoid a variety of biting insects. There is a series of products that are safe and much more effective. Burning mosquito coils have been around for 100 years and have not changed much in principle. They dissipate a cloud of fine particles containing a pyrethroid insecticide or natural pyrethrins, which kills or repels mosquitoes flying toward the cloud. Coils are available in the United States and they are very commonly used in the tropics. More recent devices heat paper mats containing a pyrethroid, which volatilizes into the immediate area. In the United States, such devices are restricted to outdoor use, but overseas electrically heated paper mats and liquid reservoirs of insecticide are commonly used indoors. A very recent product blows a small fan across a reservoir of a volatile pyrethroid in a device designed to be worn on a person. These pyrethroid-based dispersers work well as long as air currents do not dissipate the active ingredient.

Despite the complication and expense of developing new, innovative products, the public's desire for better protection from biting arthropods continues to stimulate research efforts by industry, academia, and government. Recent innovations in formulation of repellents, invention of new repellent active ingredients, scientific improvement of traps, and clever designs for area repellent devices are all encouraging signs that entomology, organic chemistry, and industry are getting closer to producing more effective products.

Professionals who give advice to the public on prevention of bites from arthropods need to evaluate the requirements of the individual and his or her activities. Knowing the health threat from the biting arthropod can influence the level of protection, effort, and expense that are justified. On a fixed property, it makes sense to remove the sources of pests as much as possible, rather than producing the pests only to have to kill or repel them. Insecticides can be used safely and effectively by the individual, but their use involves considerable skill for appropriate selection and application. If all else fails, repellents, traps, and area repellents can offer relief. They are easy for the professional to recommend because the legitimate products are safe and easy to use, but they should usually be recommended as part of a series of measures forming an integrated pest management program for the individual.

References

1. Luckmann, W. H.; Metcalf, R. L. In Introduction to Insect Pest Management, 3rd ed.; Metcalf, R. L., Luckmann, W. H., Eds.; *Environmental Science and Technology*; John Wiley & Sons, Inc.: New York, 1994, pp 1−34.
2. Mullen, G.; Durden, L. *Medical and Veterinary Entomology*; Academic Press: Amsterdam, 2002.
3. Becker, N.; Zgomba, M.; Pedric, D.; Dahl, C. *Mosquitoes and Their Control*; Springer: New York, 2003.
4. Boisvert, M.; Boisvert, J. *Biocontrol Sci. Technol.* **2000**, *10*, 517–561.
5. Braga, I. M.; Mello, C. B.; Peixoto, A. A.; Valle, D. *Mem. Inst. Oswaldo Cruz* **2005**, *100*, 435–440.
6. Strickman, D.; Frances, S. P.; Debboun, M. *Prevention of Bug Bites, Stings, and Disease*; Oxford University Press: New York, 2009.
7. Piesman, J.; Eisen, L. *Annu. Rev. Entomol.* **2008**, *53*, 323–343.
8. Ishmael, J.; MacGregor, J. A.; Manley, A. *Regul. Toxicol. Pharmacol.* **2006**, *44*, 238–248.
9. Moore, D. J.; Miller, D. M. *J. Econ. Entomol.* **2006**, *99*, 2080–2086.

10. Koehler, P. G.; Patterson, R. S. *J. Econ. Entomol.* **1991**, *84*, 917–921.
11. Moore, S. J.; Debboun, M. In *Insect Repellents: Principles, Methods, and Uses*; Debboun, M., Frances, S. P., Strickman, D., Eds.; CRC Press: Boca Raton, FL, 2007, pp 3−29.
12. Frances, S. P. In *Insect Repellents: Principles, Methods, and Uses*; Debboun, M., Frances, S. P., Strickman, D., Eds.; CRC Press: Boca Raton, FL, 2007, pp 311−325.
13. Frances, S. P. In *Insect Repellents: Principles, Methods, and Uses*; Debboun, M., Frances, S. P., Strickman, D., Eds.; CRC Press: Boca Raton, FL, 2007, pp 337−340.
14. Puccetti, G. In *Insect Repellents: Principles, Methods, and Uses*; Debboun, M., Frances, S. P., Strickman, D., Eds.; CRC Press: Boca Raton, FL, 2007, pp 353−360.
15. Strickman, D. In *Insect Repellents: Principles, Methods, and Uses*; Debboun, M., Frances, S. P., Strickman, D., Eds.; CRC Press: Boca Raton, FL, 2007, pp 347−351.
16. Moore, S. J.; Lenglet, A.; Hill, N. In *Insect Repellents: Principles, Methods, and Uses*; Debboun, M., Frances, S. P., Strickman, D., Eds.; CRC Press: Boca Raton, FL, 2007, pp 275−303.
17. Strickman, D. In *Insect Repellents: Principles, Methods, and Uses*; Debboun, M., Frances, S. P., Strickman, D., Eds.; CRC Press: Boca Raton, FL, 2007, pp 385−393.

Editors' Biographies

Dr. Gretchen E. Paluch

Dr. Gretchen E. Paluch is the Director of Basic Research at EcoSMART Technologies, Inc. She graduated from Iowa State University in 2009 with a Ph.D. in Entomology and Toxicology, with a minor in Statistics. Her appointment in the Research & Development team at EcoSMART has involved formulation, sourcing, and production aspects of newly developed contact insecticides and repellents. She has expertise in the area of insect toxicology, with an emphasis on mono- and sesquiterpenoid chemistry derived from natural sources such as plant essential oils. Her main research interest is the advancement of safe and effective botanical chemistries for use in integrated pest management systems.

Dr. Joel R. Coats

Dr. Joel R. Coats is a Distinguished Professor of Entomology and Toxicology at Iowa State University. He received his Ph.D. in Entomology, with a minor in Chemistry, from the University of Illinois, Urbana-Champaign. Dr. Coats has a diverse research background with experience in modes of action of natural insecticides and repellents, metabolism, and environmental fate of agrochemicals. His work is presented in over 180 scientific publications, and he has been recognized with the International Award for Research in Agrochemicals by the Agrochemicals Division, ACS. He is Fellow of the AAAS and the Entomological Society of America. He has trained 41 graduate students and 13 postdocs.

Indexes

Author Index

Subject Index

C

D

E

F

G

H

I

K

L

M

N

O

P

Q

R

W